About Island Press

Island Press, a nonprofit organization, publishes, markets, and distributes the most advanced thinking on the conservation of our natural resources—books about soil, land, water, forests, wildlife, and hazardous and toxic wastes. These books are practical tools used by public officials, business and industry leaders, natural resource managers, and concerned citizens working to solve both local and global resource problems.

Founded in 1978, Island Press reorganized in 1984 to meet the increasing demand for substantive books on all resource-related issues. Island Press publishes and distributes under its own imprint and offers these services to other nonprofit organizations.

Support for Island Press is provided by Apple Computer, Inc., Mary Reynolds Babcock Foundation, Geraldine R. Dodge Foundation, The Energy Foundation, The Charles Engelhard Foundation, The Ford Foundation, Glen Eagles Foundation, The George Gund Foundation, William and Flora Hewlett Foundation, The Joyce Foundation, The John D. and Catherine T. MacArthur Foundation, The Andrew W. Mellon Foundation, The Joyce Mertz-Gilmore Foundation, The New-Land Foundation, The J. N. Pew, Jr. Charitable Trust, Alida Rockefeller, The Rockefeller Brothers Fund, The Florence and John Schumann Foundation, The Tides Foundation, and individual donors.

Healthy Homes
Healthy Kids

Healthy Homes
Healthy Kids

Protecting Your Children From

Everyday Environmental Hazards

Joyce M. Schoemaker, Ph.D., and Charity Y. Vitale, Ph.D.

ISLAND PRESS

Washington, D.C. □ Covelo, California

To our children, Paul and Kimberly, and Laura and
Peter, in the hope that they and theirs will benefit
from healthy homes

© 1991 Joyce M. Schoemaker and Charity Y. Vitale

Text design by David Bullen
Illustrations by Christopher Müller

Library of Congress Cataloging-in-Publication Data
Schoemaker, Joyce M.
Healthy homes, healthy kids / Joyce M. Schoemaker and
Charity Y. Vitale.
p. cm.
Includes bibliographical references and index.
ISBN 1-55963-057-4 (cloth : alk. paper).—ISBN
1-55963-056-6 (paper : alk. paper)
1. Housing and health. I. Vitale, Charity Y. II. Title.
RA770.S44 1991
649'.4—dc20   91-21975
                                     CIP

Printed on recycled, acid-free paper

Manufactured in the United States of America

10 9 8 7 6 5 4 3 2

Contents

Acknowledgments

We wish to thank the people associated with Island Press for their enthusiasm, unflagging support, and help developing and producing this book: Barbara Dean, Barbara Youngblood, Linda Gunnarson, Connie Buchanan, Beth Beisel, Elizabeth Mook, and Will Farnam. Special thanks go to Paul Schoemaker for his numerous invaluable suggestions and criticisms throughout the course of the project, and to David Vitale for his very useful advice. Thanks also go to Mary Gochenour for her enthusiasm and help with research.

We would also like to recognize the advice and encouragement we received from the following friends and professionals: Connie Clausen, Charlotte Rancilio, Christine Rosso, Judith Stockdale, Michael Steinberg, Sandra Goldman, Ellen Paseltiner, Carla Swirsky, Linda Greenberg, Eliza Winkler and Ellen Bryant.

We have benefited from the encouragement and patience of our husbands, children, parents, and the rest of our families. The efforts of our husbands on the home front are especially appreciated.

Preface

We met sixteen years ago while teaching biology at St. Joseph's College in Philadelphia. After marrying and having children, we began to investigate the literature on environmental hazards in the home and decided to write a book that would be accessible to parents. Two years of research and writing revealed the extent to which children are vulnerable to such hazards and thoroughly convinced us of the usefulness of a book like this.

Like most parents, we wanted to provide a safe environment for our children. Like most parents, we were alarmed at the latest news about another hazard in our family's food, air or water. And like most parents, we had a sense that there was little we as individuals could do. It only added to our frustration to hear that many of the toxic chemicals in widespread use could cause cancer years after exposure. Which of us hasn't remarked in disgust, at one time or another, *"Everything* causes cancer"? But disgust wouldn't get rid of the problems. So instead of giving up and ignoring it, we decided to meet the challenge. We were determined to educate ourselves and others about hazards and the options for dealing with them. The result is this book, which offers parents information not available, to our knowledge, in any other single source.

In addition to our concern as parents, our experience as homeowners showed us firsthand what dealing with environmental hazards entails. Among the problems we faced were asbestos, radon, contaminated drinking water, and extremely-low-frequency (ELF) electromagnetic radiation from power lines. We also had to weigh the advantages and disadvantages of using pesticides on our lawn and indoors. We hope readers, instead of being overwhelmed by such problems, will come away from this book feeling that they can exercise considerable control over the environment in which their children live. Many of the solutions we propose are simple and inexpensive. The old adage "An ounce of prevention is worth a pound of cure" best sums up our message. Taking simple steps to lessen your children's exposure to environmental threats at home will reduce the risk they pose of physical and mental impairment.

Writing this book together has been an exciting and intellectually enriching experience. We feel the joint work has resulted in a book much better than anything that could have come from a solo effort—a book, we hope, that is balanced, objective, and above all, useful to parents.

Healthy Homes
Healthy Kids

Introduction

WHY READ THIS BOOK?

We spend eighteen years trying to provide a nurturing, safe, healthy environment for our children, yet despite all our concern and effort, most of us are unaware of the toxic hazards our children face at home. Some of these are radiation, from radon gas and electric blankets; contaminated air, from outgassing of synthetic building materials containing formaldehyde and from carcinogenic particles like asbestos; contaminated food and water, from pesticides and food additives; and contaminated yards, from pesticides and vehicular emissions. The aim of this book is to teach parents about such hazards and how to reduce or avoid them. Parents will learn that:

- Children, especially the very young, are more vulnerable than adults to many toxins.
- Most exposure comes from a child's daily routine at home.
- These hazards can have long-term effects like cancer, mental retardation, and asthma.
- Parents can control the environmental quality of their own homes.

WHY ARE CHILDREN SO VULNERABLE?

Children need more protection than adults from toxic hazards for two reasons: they have more exposure to these hazards, and because of their physiology, exposure puts them at greater risk. Indoors, children play close to the floor where heavier pollutants settle. Outdoors, they roll around on grass and climb trees, coming into contact with pesticides and other toxic hazards in soil. They eagerly cram potentially hazardous materials into their mouth. And they tend to eat more of certain foods like fruit that may contain toxic chemicals like pesticides.

Physiologically, children are more vulnerable because of their higher metabolic rate: they require more oxygen, and they breathe in two to three times as much air (and therefore pollutants) relative to body size than adults. Children are more physically active than adults. This also increases their breathing rate and intake of pollutants. Also, children suffer more respiratory illness: their frequently blocked nasal passages make them do more mouth breathing, which doesn't filter out particles the way nose breathing does.

Studies in humans have shown that once ingested, metals like lead and cadmium are absorbed more efficiently through the gastrointestinal tract of the young. For example, children up to age eight can absorb up to five times as much lead as adults, and they retain it longer. Studies with various species also indicate that the young are less capable of binding toxic chemicals to plasma proteins. Protein binding is important because it prevents toxic agents from reaching sites like the brain where they can do damage. Many of the detoxification systems that normally neutralize and excrete chemicals in the liver and kidneys are immature in young animals. Nor is the immune system fully functional. Humans do not build up adult levels of certain antibodies until around ten years of age.

The blood-brain barrier that protects the human brain from some toxic chemicals is not completely formed in the infant. Once inside the brain, neurotoxins—agents toxic to nervous tissue—can have devastating effects. Cells of the developing nervous system are actively growing, dividing, and migrating as well as forming complex networks. Neurotoxic chemicals can interfere with these steps, leading to permanent problems like learning disabilities. Finally, young children have a longer life expectancy during which cancers with long latency periods can develop.

WHY FOCUS ON THE HOME ENVIRONMENT?

Most of a young child's time is spent indoors. Why should we be more concerned than our parents were about the quality of the indoor environment? For one, the homes we grew up in as children were probably not as tightly sealed as most homes today. And we have introduced into these hermetic capsules a multitude of synthetic materials and household products that release toxic vapors. The result in many cases is a greater concentration of pollutants indoors than out.

Indeed, for most Americans, primary exposure to some of the most hazardous air pollutants is from indoor rather than outdoor sources. This was the conclusion of the Total Exposure Assessment Methodology (TEAM) studies conducted over the last decade by the Environmental Protection Agency (EPA). The overall finding was that even in urban areas, the concentration of organic chemicals was higher indoors than out—in some cases ten, twenty, even thirty times higher. Indoor sources are numerous; they include building materials, cigarette smoke, moth crystals, cleaning products, hot showers, and printed material.

Other forms of pollution in and around the home have become more of a threat in recent years. Hazardous materials like lead paint and asbestos, deteriorating in homes built decades ago, or manipulated during renovation work, are now contaminating household dust. Our growing use of electrical appliances and the extension of long-distance, high-voltage electricity wires have increased the extremely-low-frequency electromagnetic radiation bathing our homes and yards. This radiation is currently being researched by the EPA as a possible carcinogen. Radon, which the EPA considers the second most common cause of lung cancer after cigarette smoking, has always been a potential problem indoors. In spite of the publicity this threat has recently received, the average homeowner remains unconcerned.

Outdoor hazards are on the rise as well. The steadily expanding use of synthetic pesticides in our yards is a menace to children. More ultraviolet radiation may be striking our children now because protective ozone in the atmosphere is being depleted by chemicals. Meanwhile, close to earth, ground-level ozone (the bad ozone) regularly reaches dangerously high levels in many of our cities and suburbs.

There is also concern about the hazards in our children's food: pesticides in fruit and vegetables, chemical additives in processed food, antibiotics and hormones in milk, bacteria in poultry and meat, and industrial chemicals in fish. Most of these are the very foods we consider the healthiest for our children. Our drinking water is also at risk of contamination from an increasing diversity of toxic industrial and agricultural chemicals.

HOME HAZARDS ARE UNDERESTIMATED

In 1987, the EPA embarked on an ambitious program to identify and compare environmental problems. The idea was that in a world of limited resources, the agency should be focusing on those pollutants that pose the greatest risk to society. The conclusions of the task force of agency managers and outside experts were surprising. In addition to the obvious dangers from smokestack industries, society now faces less visible dangers, many of them caused by toxic chemicals found indoors. The task force named the following hazards as posing the greatest health risk to society (not in this order): criteria air pollutants like those that cause smog; hazardous/toxic air pollutants like benzene; indoor air pollution from radon, space heaters and gas ranges, pesticides, and cleaning solvents; drinking water contamination; application of pesticides; on-job exposure to chemicals; and stratospheric ozone depletion.

The study also revealed that not all of these hazards are now major priorities for the EPA, nor are they the hazards society finds most disturbing. One public opinion poll ranked hazardous waste sites, industrial accidents that result in the release of pollutants, and oil spills as higher risk than indoor air pollution, radon, and drinking water contamination, which came near the bottom of the list. Why the difference between the EPA experts and the public? First, the public simply does not have access to the information that the experts do. Second, studies have shown that people tend to overestimate the seriousness of hazards that are well publicized or perceived as globally catastrophic, and to underestimate the risk of more familiar hazards.

Whatever the reasons, the public is what ultimately dictates the EPA's budget and priorities, for it is the elected Congress that draws up the legislation the EPA follows. As citizens, we must become knowledgeable about the health risks from hazards like radon and lead if we want policymakers to incorporate our concerns into legislation. In the meantime, there are many things parents can do at home to minimize the risks to children. We hope this book will show the way.

CAUTION TO READERS

Some hazards examined in this book can occur in more than one place or time in the house. Parents should apply what they learn about a hazard in one chapter to other activities and locations in which the child will be exposed to the hazard.

To reduce health risks in your home, you may need to run diagnostic tests and then choose a solution. We have tried to make this process as simple as possible by giving advice on testing services and devices. Information has been gathered from many sources. Although we have tried to seek out the most reliable procedures and products, we cannot take responsibility for their results. It should also be noted that we have no financial interest in any of the products or services mentioned in these pages.

Part I

The Healthy Playroom

Children spend a great deal of their day playing at home. In most homes, there is one space that serves as a playroom. Most parents give thought to making it as safe as possible. They childproof it to remove the hazards that can result in cuts, electrocution, and poisoning. Few parents, however, examine the playroom for hazards that can cause chronic problems—respiratory disease, neurological damage, and cancer. This section of the book shows parents how to recognize and deal with lead in deteriorating paint; radioactive radon seeping through walls and floors; deteriorating asbestos; and vapor and dust from art materials.

CHAPTER I

Limiting Lead Paint
and Dust

The Problem. Paint containing lead, a toxic metal especially hazardous to the nervous system, was manufactured until the mid-1970s and is still present in homes of all types.

The Risk to Kids. Young children breathe in lead-contaminated household dust and eat lead paint chips as they crawl about. Lead damages the immature nervous system of children. Depending on the amount of exposure, it can cause a decreased IQ, learning disabilities, retardation, and death.

What to Do. Concerned parents can consult a physician about having their children tested for lead. Parents can keep the home free of lead dust and debris, get rid of old toys or furniture with deteriorating paint, remove lead paint from windows, seal or cover up lead paint on walls, and remove children from homes being renovated.

Toxic lead paint is a problem in 52 percent of American homes and in 70 percent of the houses in America's largest cities. In 1987 a Chicago child died from lead poisoning, causing outrage over the slow pace of the city's efforts to protect children from this hazard. Children who survive lead poisoning can suffer lifetime disabilities. This happened to five-year-old Desmond, who lives in Chicago and attends a special school for the handicapped. At the age of two, he had a blood test revealing a lead level four times the level considered toxic. His symptoms were weight loss, constant sickness, and an unsteady gait. After recovery, Desmond was left with permanent brain damage. Even very low levels of lead in a child's bloodstream can result in neurological damage that manifests itself in learning disabilities years later. Lead poisoning is considered the most prevalent environmental disease affecting American children. Unlike some health hazards, however, it can be eradicated.

WHY IS LEAD PAINT SO DANGEROUS?

Lead is a toxic metal that comes from harmless mineral ores. It is a versatile metal that has been in use since ancient times. The Egyptians used it in making figurines, the Romans in making roofs, pipes, and wine. In modern times lead has been added to gasoline, batteries, and paint. Lead keeps paint fresh, enhances its color, and helps it resist corrosion.

The problem with such paint is that, as time passes, it flakes and releases lead dust. With some lead paint the surface layer chalks off, becoming dust. Lead paint on windows deteriorates faster because of weathering and repeated opening and closing. Young children are likely to breathe in this dust and mouth the occasional paint chip on the floor.

Lead Poisoning

High doses of lead cause acute poisoning with wide-ranging symptoms. The U.S. Center for Disease Control (CDC) defines toxicity as a blood lead level equal to or greater than 25 micrograms of lead per deciliter blood. (A deciliter is one-tenth of a liter.) A child with this level may have a lower IQ and slowed nervous system reaction times. As blood levels climb, the symptoms grow more severe. For example, anemia and loss of nerve function in hands and feet occur at levels above 70 micrograms. Brain damage, with associated lethargy, irritability, clumsiness, tremor, coma, or even death, occurs at levels above 80 micrograms.

Low doses of lead have recently been discovered to act as a subtle but dangerous poison. Even children with levels below 25 micrograms can exhibit slowed growth and decreased IQ. Fetuses are vulnerable, since lead can pass to them through their mother's blood. Exposed fetuses show a younger age and weight at birth, and abnormalities in neurobehavioral development after birth.

There is evidence that neurological damage from low doses of lead may be permanent. The school performance of some children exposed in early childhood has been tracked by Dr. Herbert Needleman, a prominent researcher in lead poisoning. This group, tested in the first and fifth grades, and again at age nineteen, had a reading disability rate six times that of the general school-age population. The exposed children dropped out of high school at a rate seven times that of the general school-age population.

Why Are Children Vulnerable?

Children are more vulnerable to lead than adults. Children are more likely to eat and breathe lead. They spend a lot of time on the floor, where lead paint chips and dust accumulate. They like to mouth any and everything, as well as sink their teeth into painted windowsills, toys, and furniture.

Moreover, their bodies absorb more lead than do adults'—about 40 percent, as opposed to our 10 percent for adults—and they retain more of the lead they absorb. Some lead is stored in teeth and growing bones, where it can remain for decades until a stressful event like surgery triggers its release. Breastfed babies can drink lead that was previously stored in their mother's bones. Finally, because children breathe in more air than adults relative to body weight, they inhale more lead dust.

Once inside the body, lead is especially toxic to nervous tissue. This is dangerous for a young child, whose brain, spinal cord, and nerves continue developing after birth until two years of age. Brain damage during this time can be serious and permanent. A lead-poisoned child may be left with more than a lower IQ and hyperactivity; he or she may have trouble paying attention, following directions, and hearing.

Any child is at risk whose home contains paint manufactured before the federal lead paint bans enacted during the 1970s. Based on house age, some 60 to 80 percent of the units in New York, Boston, Philadelphia, and Chicago are contaminated. The U.S. Agency for Toxic Substances and Disease Registry recently estimated that about 12 million children under the age of seven are exposed to lead paint. Some 8 to 11 percent of inner-city children have lead levels over 25 micrograms. And 17 percent of pre-schoolers in urban and suburban dwellings, *regardless* of family income, have blood lead levels above 15 micrograms.

OTHER MEANS OF EXPOSURE TO LEAD

Unfortunately, children are exposed to lead from many sources. The Consumer Products Safety Commission (CPSC) effectively banned lead in household paint in the 1970s; lead was reduced to 0.5 percent in 1973, and then to 0.06 percent in 1977. But some paints, such as those used by artists and those used outdoors (the yellow lines in roads), were exempted from the ban. You should keep professional-grade paints (see chapter 4, "Getting

Rid of Toxic Art Materials") out of the hands of your children, and prevent them from chewing colored magazines or comics, which may contain lead.

Other major sources of exposure are vehicle emissions (from lead-gas combustion), soil contaminated by exterior lead paint or exhaust, food contaminated by lead cans and ceramic dinnerware, and drinking water that has been in contact with lead plumbing, including water from fountains and coolers. House carpeting can be a significant source of lead exposure to crawling and toddling youngsters. Carpeting—especially high-pile—can become a reservoir of toxic materials tracked indoors by shoes. Recent studies show that lead levels in carpets correlate with levels in the soil outside the house. Industrial lead emissions affect children living near factories. Some are exposed to lead dust brought home on the clothes of a parent employed in one of the lead-related industries, for example, lead smelters; brass foundries; battery, ceramics, or ammunition manufacturers; firing ranges; and plastics and stained-glass window producers. Airborne lead emissions may increase as landfills close and more trash is incinerated. There are also some less common sources of lead exposure: leaded-gasoline fumes, accidentally ingested shot from game birds, imported ethnic toys, home-distilled spirits, and the burning of lead-painted wood or battery casings.

A proposed Senate bill, the Lead Exposure Reduction Act of 1991, seeks to reduce lead in the environment. It would restrict the uses of certain lead-containing products and restrict the lead content in many more products, such as plastic additives, printing inks, pesticides, fertilizers, glazes, toys, fishing weights, stained glass, and wine bottle foil. It also calls for stricter labeling laws, manufacturing notices about new lead-containing products, and the recycling of lead-acid batteries. You can contact your government representatives if you support this bill.

HOW TO PROTECT YOUR CHILDREN

You can do a lot to protect your children from lead paint. Some things are simple and inexpensive, others more difficult and costly. First, keep your house clean. Regularly remove dust and debris, and watch what your children put in their mouths. To reduce the level of lead particles in indoor carpeting—a problem in homes with contaminated yard soil—have family members remove their shoes at the door. Shoe removal can reduce lead build-up in carpeting by 90 percent. Use a vacuum cleaner with power carpet beaters, which can remove five times as much dirt and dust as suction-only models. Remove any older or antique toys and furniture that your children might teethe or that might be losing paint. Second, examine your home for naturally deteriorating lead paint, and run a test to confirm the presence of lead. Next, evaluate the three main lead abatement options: replacement, encapsulation, and removal. If possible, test for lead paint and do abatement work before moving into a home. Children and pregnant women should not remain in an older home undergoing lead abatement or renovation.

How to Test for Lead Paint

You can purchase home testing kits for lead paint (see Resources). One that tests all surfaces is Lead Check Swabs. You can run four tests for about $13, eight for $16, and sixteen for $30. The test takes less than a minute. You wet a swab with a chemical solution, rub the swab on the surface, and wait for a red color indicating the presence of lead to appear. Painted wood surfaces must be prepared before testing. The layers of paint are scraped or cut away from a small spot, and the test is run on this spot.

Critics of chemical spot tests say that they depend on subjective interpretations of color, and that the presence of other metals can alter the test. Another way to test is to use a professional laboratory. Call your local health department and request a test. If it cannot be performed, someone may be able to recommend a private testing laboratory. If you decide on a private lab, you will have to provide a paint chip sample, and the test will cost about $40. Another test is to have a skilled technician use a portable X-ray-fluorescence machine. To check the level of lead dust in your home, you can hire an abatement firm to perform a surface-wipe test.

Should You Replace, Seal Off, or Remove Lead Paint?

Once you have confirmed the presence of lead paint on your walls or woodwork, you must decide what to do about it. If it is deteriorating, you should remove the paint or create an effective barrier between it and your family. Now is the time to choose among abatement options, guidelines for which have recently been offered by the National Institute of Building Sciences.

These guidelines are replacement, encapsulation, and removal. Replacement is the most expensive but least hazardous method. It is useful for lead-painted windows, doors, and woodwork.

Encapsulation or sealing off is less expensive than replacement and generally safer to perform than removal. Some encapsulants are latex coatings that, when applied, are six times thicker than paint (see Resources). The best encapsulants are impermeable to lead dust and permanent. Walls can be covered with vinyl and wood paneling. Painted floors can be covered with tile, woodwork with vinyl or new wood. Outdoor paint can be encased with siding. Information on encapsulation and encasement materials is available from lead abatement and consulting firms and from nonprofit educational organizations like the Housing Resource Center (see Resources). Critics of encapsulation say that it only postpones the lead problem, and that removal is better.

Removal creates the greatest amount of lead dust and fumes. Certain paint removal procedures should never be followed: burning off with a torch or high-temperature heat gun, sanding with power tools, grit blasting, and stripping with the toxic solvent methylene chloride (a suspected human carcinogen). The following procedures are acceptable, provided workers protect themselves and their children from the dust and fumes: heat guns and heat plates that operate below 700° F, to soften paint before scraping; chemical strippers free of methylene chloride, to soften paint before scraping; and plain dry scraping (this is a controversial method that some find too dusty). In all cases, workers should have eye protection and wear a respirator with a dust and fume cartridge and a high-efficiency particle air (HEPA) filter. Workers using chemical strippers should have an organic vapors cartridge and set up fans for ventilation, making sure all open flames (stove and furnace pilots) are turned off.

Who Should Perform Removal Procedures?

The CPSC says only professionals trained in lead abatement should do work involving a significant amount of lead dust. These people are trained to protect workers on the job, contain dust, clean up, perform clearance tests, and dispose of hazardous lead waste. One experienced national abatement contractor and consulting firm is LeadTec Services, in Baltimore (see Resources). Call your local health department for an abatement contractor near you. Contracting the work is expensive—up to $400 for an 8-by-12-foot room. Lead abatement for a complete house can run from $3,000 to $8,000.

THE HAZARD OF RENOVATION DUST

People renovating an old home or working on old furniture must protect themselves from lead dust and fumes. The CPSC recommends that workers isolate the work area, use respirators, and wear protective clothing. And they advise parents with children to get them out of the house.

One family with a Victorian farmhouse experienced this problem. Their children got lead poisoning when the house underwent major renovation. Workers sanded multiple layers of paint from floors and used torches, heat guns, and chemical stripper to remove paint from moldings on door frames. The family vacationed away from the house during much of the work and, upon their return, the parents tried to keep the children away from the work area. They hired a babysitter who kept the children occupied, mostly outdoors. Toward the end of the job the family dog, which had been regularly sitting near the workers and licking its dusty coat, became sick and was diagnosed with lead poisoning. A short time later, the mother grew weak and tired, and her five-year-old daughter began to complain of stomachaches and nausea. The girl's lead level was found to be 56 micrograms per deciliter of blood, well above the toxic threshold, and her twenty-month-old brother had a level of 87 micrograms. Both children required multiple courses of medical treatment. Elevated lead levels also were found in the parents, the babysitter, and the babysitter's two children, who had accompanied their mother on the job.

After renovation is completed, a work area isn't safe until all the dust has been carefully cleaned up. This requires vacuuming with a high-efficiency particle air machine, steam-cleaning carpets several times, and wet-mopping hard surfaces. You must also be careful in disposing of lead-contaminated debris. Call your city hall for advice on toxic substance disposal. You may be able to bring the debris, neatly wrapped and labeled, to a hazardous waste collection center in your area (see chapter 18, "Disposing of Hazardous Household Products").

POSTSCRIPT: MERCURY PAINT

Until very recently, it was legal to use mercury, another toxic metal, as a preservative in water-based latex paint. About one-third of all interior latex paint contained varying amounts of mercury. From an environmental standpoint, latex paint is preferable to oil-based paint with its toxic petroleum solvents. But mercury has turned out to be a serious threat. In 1989, during a hot July in Detroit, a family with a four-year-old son decided to paint the inside of their home. They used air-conditioning, keeping the windows closed, and the home was not very well ventilated. The job required 17 gallons of latex paint; unknown to the family, it contained mercury. At that time, manufacturers weren't legally bound to list mercury on their labels. A month after the job, the boy was diagnosed with acrodynia, a form of mercury poisoning. He had inhaled mercury fumes, which are heavier than indoor air and settle toward the floor.

The symptoms of acrodynia include rash, headache, high blood pressure, leg cramps, and muscle weakness. The Detroit boy's case was so bad he could not walk. Fortunately, after four months of hospitalization and repeated courses of therapy and rehabilitation, his legs regained their strength. Now, except for some abnormalities of the nervous and muscular systems, he is symptom-free.

In August 1990, the EPA banned mercury in indoor paint. Mercury is still present in outdoor paint, but the label must carry a warning against indoor use. Remaining stocks of indoor paint containing mercury could be sold through June 1991. For now, a parent's best defense is to ventilate well when using any paint and keep children away from fumes. Provide extra ventilation for several months after painting. Mercury fumes can linger for months. The EPA has a toll-free hotline for information on mercury in paint (see Resources).

RESOURCES

Products and Services

Baltimore City Health Department, Lead Paint Poisoning Division, 303 E. Fayette St., Baltimore, MD 21202. Telephone: (301) 396-0138; video orders (301) 396-0069.
  Offers a video on safe and effective lead paint abatement.
Baubiologie Hardware, 200 Palo Colorado Canyon Rd., Carmel, CA 93923. Telephone: (800) 441-8971 or (408) 625-4007.
  Offers a free mail-order catalogue with do-it-yourself kits for testing lead.
Certified Technologies Corporation (CERTECH), 10125 Crosstown Circle, Suite 390, Eden Prairie, MN 55344. Telephone: (800) 443-1892 or (612) 941-5093.
  Sells Certane™ 4000 Leadcoat, a latex-based lead encapsulant for walls, sills, and door frames, plus other lead paint abatement products.
Encapsulation Technologies, 310-12 North Charles St., Baltimore, MD 21201-4302. Telephone: (301) 962-5335.
  Sells Encapsulastic,™ an elastic-membrane lead paint coating.
EnviroScience Consultants, 66 Cedar St., Newington, CT 06111. Telephone: (203) 666-7167.
  Performs lead paint surveys and testing, using both laboratory and on-site analysis with X-ray fluorescent equipment.
Innovative Synthesis Corporation, 45 Lexington St., Suite 2, Newton, MA 02165. Telephone: (617) 244-9078.
  Markets a home lead paint testing kit called Lead Detective.
Lead Check, P.O. Box 1210, Framingham, MA 01701. Telephone: (800) 262-LEAD.

Has a home lead (and lead paint) testing kit called Lead Check Swabs.

Urbco, Urban Products, 50 Howe Ave., Millbury, MA 01527. Telephone: (508) 865-4103.

Sells the Insta-test Lead Paint Detector Kit, a do-it-yourself test that works on painted wood.

Sources of Information

Center for Disease Control (CDC), Lead Poisoning Prevention Branch, 1600 Clifton Road, MS: F-28, Atlanta, GA 30333. Telephone (408) 448–4880.

Contact them for the new health guidelines for children with low but elevated blood lead levels.

Consumer Product Safety Commission, Consumer Information Center, Dept. 467X, Pueblo, CO 81009. Telephone: (800) 638-2772.

Will send you the booklet "What You Should Know About Lead-Based Paint In Your Home." Send your name and address with 50¢.

Environmental Health Watch and Housing Resource Center, 4115 Bridge Ave., Cleveland, OH 44113. Telephone: (216) 961-4646.

Offers information on encapsulation materials for lead abatement in houses.

EPA. Mercury hotline: (800) 858-7378.

Tells you if mercury is contained in the brand of paint you have purchased.

Housing Resource Center, 1820 W. 48 Street, Cleveland, OH 44102. Telephone: (216) 281-HOME.

Offers the "House Mending Notebook," with individually available pages devoted to home repairs and hazards.

LeadTec Services, 522 Beck Ave., Baltimore, MD 21221. Telephone: (301) 682-5323.

Publishes a newsletter, *The Baltimore Leadletter*, with information on methods, research, guidelines, and government regulations. Contact them for a complimentary copy.

CHAPTER 2

Ridding Your Home
of Radon

The Problem. Radon is an invisible radioactive gas that seeps from the ground into homes.

The Risk to Kids. Children, whose bodies are more vulnerable to radiation, can receive radiation exposure equal to thousands of chest X-rays per year in a house with high radon levels.

What to Do. Everyone should run one of the easy, inexpensive radon tests and, if necessary, reduce radon levels by ventilation or prevention of radon's entry into the home.

One day in 1984, Stanley Watras arrived at his job at the Limerick Nuclear Power Plant in Pottstown, Pennsylvania, and walked through the radiation detector. It triggered, and so did our national awareness of radon. Ironically, the source of his contamination was not the place he worked but his home. It registered extremely high levels of the radioactive gas radon. In fact, he and his family were exposed to a level of radiation equal to 455,000 chest X-rays per year!

Today radon is considered one of the nation's most serious environmental health hazards. The gas is invisible and odorless, a silent killer. According to radon expert Anthony Nero, the hundreds of thousands of Americans who live in homes with high radon levels receive as much radiation exposure in a year as those people who lived near Chernobyl received during the course of 1986, the year the nuclear plant there exploded.

WHAT IS RADON AND HOW CAN IT HARM YOUR CHILDREN?

Radon is created during the decay or breakdown of atoms of uranium, which are naturally found in the earth's crust. When we breathe radon gas, its radioactive decay products, called radon daughters, are taken into our lungs. Radon daughters are chemically reactive particles that "stick" to lung tissue. Alpha radiation is released from decaying radon daughters. Like other forms of ionizing radiation, alpha radiation can change the structure of DNA, the vital genetic information in all our cells. This change or mutation can lead to lung cancer.

The EPA estimates that radon causes between 5,000 and 20,000 of the approximately 140,000 lung cancer deaths per year in the United States. This makes radon the second largest cause of the disease after cigarette smoking, which is responsible for about 116,000 deaths.

Children Are at Greater Risk Than Adults

Children tend to spend more time at home and in basement playrooms than adults, thus making their exposure greater. Moreover, children, because of their higher metabolic rate, inhale two to three times as much air (and radon) relative to their size as adults. They are less efficient at clearing particles like radon daughters from their airways, which are narrower than adults'.

Although there is no direct evidence that radon harms children more than adults, there is ample evidence that children are more vulnerable to other sources of ionizing radiation. This has been shown in studies on cancer rates following exposure to nuclear fallout, as well as results from exposure of children to X-rays. One estimate is that children are five to ten times more vulnerable to radiation than adults. Children's high rate of cell proliferation may facilitate the development of cancer: the more cells divide, the more opportunity for cancer-causing mutations.

RADON HOTSPOTS

Places that have brought radon to the attention of the American public are the uranium-mining areas of Colorado, those sections of Florida where phosphate, mixed with uranium, is mined, and that part of the eastern United States known as the Reading Prong, a geological formation of soft granite rocks rich in uranium running from eastern Pennsylvania to

Connecticut. These areas, it was discovered, had many homes with high radon levels. However, there were some surprises. For example, even though the Watrases' home (in the Prong) had one of the highest radon levels ever reported, the radon level in the next door neighbor's home was less than the national average. This was one of the first indications to scientists and the public that radon levels in homes cannot be easily predicted simply based on where the home is located.

In addition to a rich deposit of uranium in the ground under your home, the overlying rock and soil must have cracks and pores that allow radon to reach your home's foundation. Grainy, loose soil allows easier passage of radon; heavy soils like clay tend to seal it. So even if an area has a high concentration of granite or shale rock rich in uranium, homes may not necessarily have high radon levels.

HOW RADON GETS INSIDE HOMES

Radon gas seeps up through cracks and pores in the ground and enters the foundation of a home. Typical entry routes are cracks in the concrete slab on which a home rests; pores, cracks, and mortar joints in the concrete walls of a basement; joints between the basement floor and walls; exposed soil in a sump pump or crawl space; loose-fitting pipe holes in the basement floor and walls; the open tops of basement block walls; and building materials such as stone. Tap water may also be a source of radon in a home supplied by a private well or small community water system. See chapter 12, "Curbing Contaminants in Water."

Homes can literally suck radon up, because air pressure inside is generally lower than air pressure in soil. Whenever you turn on an exhaust fan or use your chimney, you lower the air pressure of your home. And when warm indoor air rises and escapes through small openings in your home, it creates a partial vacuum, lowering air pressure.

The highest radon levels occur in the basement, or on the first floor of a home built on a concrete slab. These areas have the most contact with the ground. Since many playrooms are located in renovated basements or on first floors, controlling radon levels is a top priority.

Are there any home characteristics that contribute to high radon levels? According to Bernard Cohen, author of *Radon: A Homeowner's Guide to Detection and Control*, features like construction materials and the age of your house have surprisingly little effect on radon levels. Weatherizing, which would be expected to raise radon levels significantly, is apparently responsible for an increase of only about 10 percent.

TESTING FOR RADON

The only sure way to know if your home has too much radon is to test for it. For this reason, and also because radon testing is easy and inexpensive, the EPA is advising that all homes be tested.

The Screening Test

Radon testing is simple and inexpensive. For the first test, called a screening test, there are several radon detectors on the market. These are passive devices that are exposed to your indoor air for a period of time and then sent to a lab for analysis. The most commonly used are the activated-charcoal detector, the alpha-track detector, and the electret-ion chamber. You can obtain detectors from a variety of places, including your local hardware store and radon testing companies. Testing with detectors costs between $15 and $25, which includes laboratory analysis of the data.

Detectors are usually placed in the basement, where radon levels are expected to be highest, especially if it is used as a playroom, family room, or bedroom. Additional detectors can be put in other frequently used rooms. The activated-charcoal test must be run for three to seven days. Alpha-track and electret-ion tests can be short term (two to seven days) or long term (one to twelve months). The advantage of long-term tests is a more reliable answer: since radon levels fluctuate from day to day and season to season, the longer the testing period, the better the information. Short-term tests have the advantage of letting you know in a matter of days whether your house has a radon level that may endanger your family's health if action is not taken quickly.

When you want a fast answer, do a short-term test under worst-case conditions: place the detector in the lowest level of your home with windows and doors closed and fans turned off. This strategy will measure the highest radon levels possible in your home. If, under these conditions, you get a relatively low reading, under normal, open-house conditions your radon level will be even lower. You should also conduct a screening test during winter months, when your house is closed up. Radon levels tend to be higher in winter than in summer, probably because the furnace is drawing air from the basement and lowering indoor air pressure.

A Fast Test for Home Buyers

Continuous monitoring devices can determine radon levels in a matter of hours. For setup and operation, you will need a trained technician from a radon testing company. These devices can be used when you are purchasing a home and want to know the radon level before you sign on the dotted line. An important advantage of monitoring devices is that the homeowner cannot control the results. Major disadvantages are the short testing period, which gives less reliable results, and the expense—$100 to $300.

Radon Testing Companies

You may be wondering how to contact a reputable radon testing company. Radon detection is not government regulated. As in any industry, there may be unscrupulous or incompetent practitioners. To help the consumer avoid them, the EPA developed the National Radon Measurement Proficiency Program (see Resources), which publishes lists of companies whose testing methods have been evaluated by the EPA. The EPA does not certify or endorse the companies. However, they have met certain requirements regarding adequacy and reliability. Test kits purchased in a store should indicate that they have been evaluated as part of this program.

Interpreting Test Results

The concentration of radon in homes is measured in picocuries per liter of air (pCi/l). A pico is one-trillionth of a curie, a standard unit for measuring radiation. Outdoor (background) concentrations of radon tend to run around 0.2 pCi/l. Concentrations in homes typically range from 1 to 3 pCi/l. Higher concentrations, in the range between 10 and 100 pCi/l, are found in homes throughout the country. The Watrases, mentioned at the start of this chapter, had the astounding concentration of over 2,000 pCi/l in their living room, one of the highest levels ever measured inside a home.

Follow-Up Testing

If a short-term screening test reveals a radon level above 4 pCi/l, the EPA recommends doing a follow-up test before taking major, perhaps costly, steps to lower the radon level in a home. The follow-up test is run over a longer period of time to obtain a more accurate picture of the radon level. It should be run in one or two of those rooms that are most used by a family. This would include the basement if it is used as a living space.

Generally, the higher the screening test result, the shorter the follow-up test should be, thus minimizing a family's exposure to radon. For example, if your screening result is between 4 and 20 pCi/l, the detectors (alpha track or electret ion) should be exposed for about one year. In the unfortunate case that the result is greater than 200 pCi/l, follow-up measurement should run no longer than a week. A helpful booklet describing the details of follow-up testing is the EPA's "Citizen's Guide to Radon" (see Resources).

WHEN TO LOWER RADON LEVELS

The EPA considers levels in the range of 4 to 20 pCi/l to be above average, and recommends taking action to lower them. The EPA's "action level" of 4 pCi/l, although a widely followed guideline, is controversial among scientists and regulators. Other countries have set different action levels. Canada's is five times higher; the United Kingdom's is 10 pCi/l for existing buildings and 2.5 pCi/l for new construction.

The Controversy over Action

While the link between high radon exposure and lung cancer is unquestionable, the risk from low exposures is uncertain. To predict lung cancer rates for the general public, cancer rates in uranium miners were used. However, because the environment of miners is more polluted than that of the average homeowner, use of this data is questionable. Also, there is a problem extrapolating from the effect on miners of high radon exposure to determine the effect on the general public of low exposure. Extrapolation assumes that the risk of cancer is proportional to the amount of exposure. However, this has never been proven for low exposure, the damage from which the body may be able to repair.

In spite of the controversy, one must keep in mind the enormity of the health risk from radon as compared with other environmental risks. The estimated risk of dying from radon-induced lung cancer for the average American living in a home with a 1.5 pCi/l radon level is 0.4 percent. This dwarfs the risks of many outdoor pollutants like pesticides and benzene, which are regulated by government to keep the risk of premature death below 0.001 percent. Considering this and the vulnerability of children, it would be prudent to follow the EPA's guidelines, even if they are not perfect.

When to Act

If your radon level is greater than the EPA action level of 4 pCi/l, you should reduce it according to the following timetable:

- 4 to 20 pCi/l within a year or two.
- 20 to 200 pCi/l: within several months.
- Greater than 200 pCi/l: within several weeks. If this is not feasible, increased ventilation or temporary relocation should be considered until radon levels are brought down.

These are general guidelines. You should also consider your family's lifestyle. How long have you lived in your home, how much longer do you expect to live there, and how much of the day do you actually spend there? The longer your exposure, the greater the risk of lung cancer. Do any of the family members smoke? Smoking increases the risk of developing radon-related lung cancer by a factor of ten. Passive smoke inhalation by young children may put them at greater risk from radon (see chapter 13, "Reducing Harmful Gases," for a discussion of passive inhalation). Do your children use the basement, where radon levels are highest, as a playroom or bedroom?

In sum, the greater your family's actual exposure to radon, the more conservative you may want to be in following the action timetable. For example, if a radon test indicates a level of 20 pCi/l in your basement and your children use this area as a playroom, you may want to lower the radon level as soon as possible rather than waiting a year.

RIDDING YOUR HOME OF RADON

You have two basic options for lowering the radon level in your home: increasing ventilation, or preventing radon from entering in the first place. Solutions range from the simple and inexpensive to the intricate and costly. Depending on the severity of your problem, you may want to use several methods. The following suggestions are made to help you work out a practical approach.

Make an analysis of the air flow through your home. Ventilation strategies for diluting radon are most effective in newer homes, which are more tightly sealed. If your home is older, chances are there is already a lot of natural ventilation around window and door frames. Increasing ventilation in an old home may not substantially decrease radon levels.

There are several ventilation strategies you can try if your home is tightly built. Open lower-level windows and doors. Install fans for a more controlled influx of air, or a heat-recovery ventilator in extreme climates. The latter is an energy-efficient device that uses air-conditioned or heated indoor air to cool or warm fresh outdoor air as it is drawn into the home.

Is it practical to open windows in your home? The answer is no in extremely cold or warm weather, unless you can close off the ventilated area (usually the basement) from the rest of the house. If your children play or sleep in the basement, opening windows during inclement weather would not be practical. Heat-recovery ventilation (HRV) may be a cost-effective solution. The initial price, between $800 and $2,500 installed, may be offset by sizable savings on heating and cooling. The EPA advises that HRV alone can be effective in lowering the radon level to below 4 pCi/l when it is no more than about 10 pCi/l to start with.

Don't rely on air cleaners to rid your home of radon. Many modern homes have air cleaners that are either hooked up to the heating or ventilation system or used as portable units in the living space. These devices remove small particles from air by means of filters or electrostatic precipitation. They have not been demonstrated to be effective in reducing the health risks of radon.

Are you unintentionally depressurizing your home? Because combustion appliances like wood stoves, fireplaces, furnaces, and hot water heaters can lower indoor air pressure, you should make sure you have a source of outside air when these units are operating. An open window or duct that runs from an outside wall to the combustion unit will suffice. If exhaust fans are used in the kitchen and bathrooms and near a clothes dryer, open a window to replace the air.

See whether your tap water is a source of airborne radon. This is an important consideration for people with private wells or for those who draw from a small community water supply. See chapter 12, "Curbing Contaminants in Water," for more details about this source of radon.

Make a visual inspection of your home for places where radon can enter. Some favorite entry points are the exposed earth in crawl spaces, openings around utility pipes and sump pumps, and cracks in basement walls and floor. You can create a barrier by covering your crawl space with concrete and filling openings with a gas-proof, nonshrinking sealant like polyurethane caulking. Floor drains exposed to the earth should be fitted with water traps, and sumps should be capped and sealed. Keep in mind that sealing and covering are only temporary—as a house ages and settles, new cracks will appear in the foundation—and that complete sealing is impractical if not impossible. Still, according to the EPA, sealing most successfully reduces radon levels to below 4 pCi/l when initial levels are moderate, less than about 20 pCi/l.

Retesting is crucial. After trying one or a combination of the above strategies, test again for radon. If the level has not substantially come down, you may need a more sophisticated strategy like subslab suction and block-wall suction, in which air containing radon is pumped from under and around your home's foundation before it can enter your home. Suction also lowers air pressure outside your home, helping remove radon. Since suction is costly ($800 to $2,000 installed) and requires an experienced contractor or homeowner, it is recommended only when radon levels are high (greater than 20 pCi/l). Unfortunately, suction means that you will have unsightly pipes running from floor to ceiling or walls to ceiling, a consideration if your basement is used as living space. The EPA booklet "Radon Reduction Methods: A Homeowner's Guide" has information on suction methods (see Resources).

What can you afford to spend? You will want to choose a cost-appropriate strategy. For a radon level below 4 pCi/l, a costly method such as subslab suction is unnecessary. However, for a more severe radon problem, this combined with other methods such as sealing cracks may be necessary and well worth the expense.

Hire a reputable radon contractor. For advice, contact your state radiation protection or EPA office, local building trade associations, realtor associations, or chamber of commerce. Your state or regional EPA office or your state radiation protection office will send you a pamphlet on the National Radon Contractor Proficiency Program, which lists proficient contractors. The office will also mail you a variety of other helpful booklets on radon (see Resources).

RESOURCES

Sources of Information

EPA. (See appendix for regional offices.)

> Provides the following free booklets: "A Citizen's Guide to Radon: What It Is and What to Do about It"; "The National Radon Measurement Proficiency Program," a list of companies in your state experienced in measuring radon; "The National Radon Contractor Proficiency Program," a list of contractors in your state experienced in reducing radon levels; "Radon Reduction in New Construction: An Interim Guide"; "Radon Reduction Methods: A Home-owner's Guide"; "Removal of Radon from Household Water"; and "Radon Reduction Techniques for Detached Houses: Technical Guidance," for the professional radon reduction contractor or the homeowner.

Radon: A Homeowner's Guide to Detection and Control. Bernard Cohen. Mt. Vernon, NY: Consumer Reports Books, 1987.

Radon: The Invisible Threat. Michael Lafavore. Emmaus, PA: Rodale Press, 1987.

CHAPTER 3

Controlling Crumbling Asbestos

The Problem. Asbestos, a durable mineral used as an insulator and fire retardant in homes built between 1920 and 1970, can—if disturbed or deteriorating—release tiny carcinogenic fibers into the air.
The Risk to Kids. Children, who relative to their size breathe in more pollutants than adults, may develop cancer up to several decades after exposure if they inhale enough fibers from disturbed or deteriorating asbestos.
What to Do. Asbestos should be identified and professionally removed from the home if it is likely to release fibers as the result of decomposition or manipulation during renovation.

In August 1988, the Selph family hired workmen to install a new heating and air-conditioning system for their Florida home. When they cut into wallboard to make air ducts, Mrs. Selph began worrying about the gray particles powdering the interior of the house. A week later, tests confirmed that the particles were asbestos fibers. Unknown to the Selphs, their home contained wallboard composed largely of a short, curly asbestos fiber known as chrysotile. The Selphs and their three boys vacated the house immediately, leaving behind almost everything they owned. They began living in a rented house and wearing donated clothing. They worried about their contaminated house—a contractor wanted $17,000 to remove the wallboard, a figure the Selphs could not afford. They also wondered what health problems they might face someday—perhaps many years later—because of their exposure to asbestos.

WHAT IS ASBESTOS?

Asbestos is the name for various silicate minerals whose fibers are not easily broken down by natural processes. Asbestos is thus an excellent strengthening, insulating, and fireproofing agent. It has also been used for acoustical insulation and decorating surfaces. Millions of tons of asbestos was added to the environment over the course of almost a century. Asbestos is often a light gray color, but it can take on various appearances. For example, the pipe insulation can look like corrugated cardboard. Hot air ducts are commonly wrapped in grayish-colored paper containing asbestos. Airborne asbestos fibers are practically invisible.

Common Locations

Asbestos was used in about a quarter of all houses and apartment buildings constructed or remodeled between 1920 and 1970. Some asbestos-containing products continued to be manufactured beyond 1970. Hairdryers with asbestos head shields were not recalled until 1979. Some asbestos pipe insulation was in production until 1972. Ceiling coating and textured paint containing asbestos were sold until 1978.

Below is a list of places where asbestos is commonly found:

• Vinyl floor tiles, linoleum flooring, and adhesives
• Ceilings (sprayed-on soundproofing, patching compounds, and textured paints)
• Wall, pipe, and furnace ducts (insulation)
• Appliances
• Stoves and furnaces (on door gaskets, and in insulation)
• Roofing and siding shingles

In general, many interior asbestos materials are friable—that is, they crumble easily—while exterior materials like roofs, shingles, and siding are not.

The automobile—eroding brake linings, clutches, and transmissions—is another major source of asbestos. A helpful booklet, "Asbestos in the Home," jointly produced by the Consumer Product Safety Commission (CPSC) and the EPA, gives further details about where asbestos can be found and when certain asbestos products were banned (see Resources). A gradual ban of nearly all asbestos products in the United States began in 1986 and will culminate in 1996. In 1986, the CPSC required all consumer products containing asbestos to be labeled as such.

WHY IS ASBESTOS DANGEROUS?

Undisturbed and intact, asbestos is harmless. Danger arises only when it releases fibers in the air. Once inhaled, they lodge in internal tissues. Exposure can result in three serious disorders: asbestosis, an irreversible scarring of the lung tissue; lung cancer, which frequently occurs twenty to forty years after initial exposure; and malignant mesothelioma, a cancer of the tissues lining the chest and intestinal cavities. Asbestos, in addition to being a carcinogen, can also adsorb—or carry on its surface—carcinogenic chemicals from the air and carry them into the lungs.

Repeated exposure to asbestos increases risk. There are an estimated 10,000 deaths per year as a result of past exposure to asbestos. Children are especially vulnerable because of their higher breathing rate, which makes them vulnerable to more air-borne pollutants. Children are less able to clear pollutants from the upper and lower respiratory tract. They also breathe more often through the mouth—their nasal passages blocked because of colds—giving inhaled particles a direct route to the lungs.

ARE CHILDREN REALLY AT RISK TODAY?

In the past, children of asbestos workers got sick from exposure to fibers brought home on a parent's work clothes. Today exposure for children comes mainly from asbestos products and construction materials. The danger depends on the amount of asbestos fiber being released in the air.

Ideally, children should not breathe asbestos fibers at all. Medical experts have pronounced that no level of exposure is safe. The EPA has taken a cautious stand, asking schools to evaluate asbestos-containing materials with an eye toward abatement (encasing or removal). Abatement is expensive and has prompted criticism. Some scientists feel two factors make school asbestos materials nonthreatening. First, the level of airborne asbestos fiber in schools and buildings is generally a small fraction of the level currently permitted in the U.S. workplace. Second, the most common type of asbestos used in U.S. schools and buildings, chrysotile, is now considered by some to be less dangerous than other kinds of asbestos. Asbestos fibers are divided by shape into two groups: curly ("serpentine") and needlelike ("amphibole"). Chrysotile, which is serpentine, is not thought to penetrate lung tissue as deeply as amphibole asbestos like crocidolite and amosite.

But there is still need for caution at home. First, a parent cannot take for granted that chrysotile is the only asbestos in the home. Unlike most other industrialized nations, the United States never regulated the use of the dangerous amphibole fibers. And these fibers sometimes contaminate commercial deposits of chrysotile asbestos. Second, a high concentration of airborne fibers presents a health risk no matter what kind of asbestos is present. And certain activities at home such as removing asbestos floor tiles and sanding underlying asbestos-containing cement can suddenly and substantially raise the concentration.

HOW TO DEAL WITH ASBESTOS

The first step is to identify asbestos in the home. Once this has been done, you can evaluate whether to leave it alone or remove it. Professional services are highly recommended for certain procedures.

If you are concerned about a suspicious appliance, call the manufacturer and ask. Give the model number and age of the product. In the case of an untraceable older product, the CPSC suggests for safety's sake that you assume it contains asbestos. With regard to building materials, the EPA recommends that a qualified professional survey your home. This will cost $300 to $500 but is well worth the expense. The advantages of using professionals are several: they know where to look for asbestos, they take samples properly, and they can offer advice about how to proceed. Getting a survey done before you remodel or buy a new home is also recommended.

No one can visually distinguish chrysotile asbestos from the more dangerous type. A laboratory will differentiate fiber type in bulk samples for about $35. To find an accredited, reliable lab, contact the National Institute of Standards and Technology (see Resources).

Remember not to disturb suspicious material yourself; you might increase the level of airborne fibers. Don't try to rip out asbestos insulation, and don't sand floor tiles or backings that might contain asbestos. If you vacuum asbestos dust, the fibers will pass through the vacuum bag and disperse through the air. If a pile of asbestos dust has collected under a pipe, wet it down with water before removal.

Should You Seal Off or Remove Asbestos?

Things don't look so simple at this point. You are dealing with tiny, perhaps invisible fibers that might drift into the air the minute you manipulate a suspicious material. Furthermore, you will have to spend some money and carefully choose someone to survey your home. Is there no silver lining to this cloud?

Well, maybe. In some cases, the best thing is to leave asbestos alone. Only remove asbestos material if it is deteriorating (broken open and crumbling), in a frequently used area, or affected by renovation. Be sure that any material you leave alone is well encased. Seal small areas where asbestos insulation is open to the air with special tape available in hardware stores. Lay new floor tiles over old asbestos tiles. Remember, asbestos is harmless unless it is releasing airborne fibers.

If circumstances indicate the need for asbestos removal, you will need a contractor who is certified, experienced, and able to provide references. Today nearly every state requires asbestos workers who work in schools or public buildings to be trained or licensed. State environmental agencies can help you locate certified asbestos contractors. State departments of health have current lists, and Asbestos Coordinators in state EPA offices have knowledge of the work records of area contractors. The CPSC-EPA booklet "Asbestos in the Home" has general guidelines for dealing with contractors.

Removal involves sealing off the work area and restricting access to workers. Workers wear approved respirators and protective clothing. The job may last several days. Afterwards an air monitoring test must be run to see if all asbestos is gone.

IS FIBERGLASS OR OTHER INSULATION SAFE?

Builders have had to find alternatives to asbestos insulation, and it is natural to wonder how safe these are. Fiberglass, which has replaced asbestos in much home insulation, carries its own risks. Epidemiologic studies have shown an increased incidence of lung cancer in workers exposed to uncontrolled levels of fibrous glass dust. The Occupational and Safety Health Administration (OSHA) requires industry to add warning labels to carcinogenic products, but the fiberglass industry is not yet in full compliance. Children should not be exposed to any airborne fiberglass fibers. These include loose fibers blown into attics as insulation, and any fibers inadvertently released by the manipulation or deterioration of a fiberglass household product. Today fiberglass dust is suppressed in the workplace, and workers wear protective eyewear and respirators, and wash their clothes separately.

The consumer should gather as much information as possible before buying insulation. A number of such materials release toxic gas or irritating particles. These include urea-formaldehyde, polyurethane, fiberglass on Kraft paper with asphalt adhesive, cellulose, vermiculite, and perlite. The environmental watchdog group Greenpeace reports that the safest insulation today is rock wool or aluminum-backed fiberglass. Other sources recommend cork and Air Krete as nontoxic (see Resources). Cork is the vermin-resistant outer bark of an evergreen tree. Air Krete is an inorganic, cement-related but lightweight product containing magnesium oxide. It is fireproof and almost completely odor free.

RESOURCES

Products and Services

Air-Krete, P.O. Box 380, Weedsport, NY 13166. Telephone: (315) 834-6609.
    A nontoxic, ultralight but cementlike thermal and acoustical insulation product.

Sources of Information

Asbestos Ombudsman. Telephone: (800) 368-5888 and (202) 557-1938.
    Responds to questions and concerns about asbestos in schools.
Center for Environmental Management, Tufts University, Curtis Hall, 474 Boston Ave., Medford, MA 02155. Telephone: (617) 381-3531.
    Offers information on asbestos testing and abatement for homeowners.
EPA. TSCA assistance line: (202) 554-1404.
    Gives the names of qualified asbestos testing laboratories and training programs in your area.
The Healthy House: How to Buy One, How to Cure a Sick One, How to Build One, John Bower, New York: Lyle Stuart, 1989.
    Discusses alternative home insulation materials and their health effects.
National Voluntary Laboratory Accreditation Program, National Institutes of Standards and Technology, Building 411, Rm. A124, Gaithersburg, MD 20899. Telephone: (301) 975-4016.
    Send a self-addressed envelope to this address for a list of certified asbestos testing and analysis laboratories.
Publication Request, Consumer Products Safety Commission, Washington, D.C. 20207. Telephone: (800) 638-CPSC.
    Write this address for the CPSC-EPA booklet "Asbestos in Your Home."
Victims of Fiberglass, Box 440, Meadow Vista, CA 95722. Telephone (916) 878–7748.
    Offers information on the hazards of fiberglass.

CHAPTER 4

Getting Rid of Toxic Art Materials

The Problem. Working with art materials, especially those made for professional artists, can expose children to extremely dangerous substances like lead and organic solvents.

The Risk to Kids. Long-term exposure to these substances puts children at increased risk for neurological disorders and chronic diseases like cancer.

What to Do. Provide your children with nontoxic products like those that carry the AP or CP label from the Art and Craft Materials Institute (ACMI). If you have an art studio at home, it should be isolated from the rest of the house and off-limits to your children.

Arts and crafts is one of the most enjoyable activities of childhood, but long-term exposure to certain compounds in materials like paint can cause neurological problems, organ damage, and cancer. Professional materials are of particular concern. This is illustrated in the case of a couple, stained-glass workers, that used their kitchen as a makeshift art studio. Their eighteen-month-old daughter often played in the kitchen while work was in progress. A blood test later revealed that she had a dangerously high lead level, one associated with learning deficiencies. She had inhaled lead in dust or soldering fumes released while her parents worked.

THE HAZARDOUS MATERIALS

There are six main classes of toxic art materials: metals, solvents, pigments and dyes, acids and alkalis, silica and talc, and aerosol components.

Metals such as lead, cadmium, chromium, and mercury are found in inorganic pigments used in some paints and glazes. Metals are also used for metal-casting, jewelry-making, welding, soldering, and stained-glass work. Artists are exposed to dust from dry pigments and to fumes generated during welding or soldering. The health effects of lead exposure are discussed in chapter 1. Lead is still a component of many paints used by professional artists, even though it was lowered to negligible levels in house paints in the 1970s. Cadmium and chromium compounds are associated with increased risk of lung and prostate cancer. They may also cause kidney and lung damage as well as birth defects.

Solvents like turpentine are used as oil paint and varnish thinners and as hand and brush cleaners. Most can irritate the skin, eyes, nose, or throat, leading to headaches, nausea, and loss of coordination. Repeated exposure to solvents can cause severe skin reactions, allergies, even coma. One tablespoon of turpentine can kill a child if swallowed.

Some solvents, like benzene, now banned, are carcinogenic. Others (styrene and toluene) can damage internal organs like the liver and kidneys.

Pigments and dyes are made mostly from petrochemicals and have not been adequately tested for their health effects. Pigments are used in paints. Dyes are used in felt markers and colored inks.

Acids and alkalis are used to etch glass, to clean and etch metals, and in photography. Vapors can burn the eyes, nose, and lungs. Diluted acids and alkalis can cause skin irritation or sensitization.

Silica and talc are components of clays and glazes used in ceramics. Repeated inhalation of silica dust by potters handling dry clay powders or glazes can cause a severe respiratory disease marked by shortness of breath and scarring of lung tissue. Silica is also found in certain sculpting stones like quartz, sandstone, or granite. Talc, a component of many low-fire clays, is dangerous because it can be contaminated with asbestos. Inhalation of asbestos-containing talc dust over a period of time may cause mesothelioma (cancer of the chest lining) and asbestosis, a form of lung scarring.

Aerosol paints, varnishes, and adhesives deliver a fine mist of their toxic components that can linger in the air for hours. Toxic propellants and solvents used in sprays pose an additional hazard to health and the environment.

WHY ARE CHILDREN VULNERABLE?

Such toxic materials can cause both acute and chronic health problems. Acute effects develop rapidly, sometimes when a product is being used for the first time. They include accidental poisoning from ingestion, pneumonia from inhalation, and temporary skin irritation. Chronic effects become apparent after long-term exposure to a substance. They range from skin allergy and organ damage to cancer and neurobehavioral disorders like hyperactivity. Chronic effects are silent and insidious; a medical problem may not be realized until damage to the body is irreparable.

Several factors make children more vulnerable to toxic compounds in art materials. By nature inquisitive, children are more likely to ingest these substances. Attractive colors and textures tempt them to perform taste, smell, and smear tests.

Children are also physiologically more vulnerable. Relative to their body weight, they inhale more air and therefore more pollutants than adults. Also, they have about two and a half times the skin surface area relative to body weight as adults, giving them greater exposure to chemicals absorbed through the skin. Children have fewer defenses against pollutants. Their skin is thinner and more susceptible to chemical burns from products like solvents, acids, and alkalis. Their immature immune systems, livers, and kidneys may make them less able to neutralize or detoxify such substances. Also, exposure of developing organs to toxic compounds can lead to irreparable damage. This is evidenced by the subtle learning impairment that results from exposure to low levels of lead. Unborn children are particularly at risk, because heavy metals like lead and mercury can cross the placenta and harm the embryo.

HOW ARE CHILDREN EXPOSED?

Children are exposed when they use materials themselves or when their parents use them. Unborn fetuses are passively exposed in the womb, breast-feeding newborns through their mother's milk.

There are few well-documented cases of children becoming either acutely or chronically ill from exposure to art materials. This is partly because of an absence of long-term medical data. Much of our knowledge of chronic health problems associated with certain chemicals comes from studies of industrial workers exposed on the job. For example, the carcinogen benzene was banned from use in many art products as a result of its effects on industrial workers. Similar health data for artists is now becoming available. A National Cancer Institute study showed that among male artists the death rate from heart disease, leukemia, and cancer of the brain, kidney, bladder, colon, and rectum was significantly greater than the rate for the general population.

Children presumably face a much smaller risk then professional artists. However, you may want to err on the side of caution when dealing with known carcinogens. In most cases, what constitutes safe levels of contact is unknown.

WARNING ABOUT HAZARDS

Federal regulations should warn us about dangerous art materials. Do they? The Federal Hazardous Substances Act of 1960 requires labeling of art products that produce immediate sickness or death. The product label must include the word DANGER, WARNING, or CAUTION (DANGER indicating the most severe hazard), the name of any hazardous ingredient, the type of hazard, precautions to take, and first-aid suggestions.

Since the 1940s, the ACMI, a nonprofit organization of manufacturers of art materials, has voluntarily labeled many art products for chronic health hazards. About 85 to 90 percent of art manufacturers, including the American Crayon Company, Binney and Smith, and Sargeant Art, are members.

The ACMI gives three nontoxic labels to products that have been evaluated by the institute's toxicologist and have been determined to "contain no materials in sufficient quantities to be toxic or injurious to humans or to cause acute or chronic health problems." These labels are AP (Approved Product) Nontoxic, CP (Certified Product) Nontoxic, and Health Label Nontoxic. These products are safe even for preschoolers whether inhaled, ingested, or absorbed through the skin. The CP seal also signifies that a product meets specific performance standards. Art materials that contain toxic substances and are appropriate only for adults and children over twelve years of age receive the Health Label seal with cautions and instructions for use. For some fifty years, in the absence of federal legislation requiring such labeling, parents and schools have relied on ACMI's nontoxic seals to assure them of the safety of art materials. You can obtain a list of these products by writing the ACMI (see Resources).

The ACMI labels are still in use today on many products, although federal law now mandates uniform labeling of art products that may cause chronic health problems. The 1988 Labeling of Hazardous Art Materials Act (Public Law 100-695) directs manufacturers to have their products evaluated by a toxicologist under the provisions of ASTM D-4236, the standard for toxicological evaluation that ACMI had been using voluntarily. The new law requires manufacturers, as well as packers and importers, to supply the following information on their labels: a warning about possible chronic health effects, hazardous ingredients, guidelines for safe use, a warning that the product is unsuitable for children, and the phrase "Conforms to ASTM D-4236." Products that have not been determined to pose a chronic health hazard are labeled only "Conforms to ASTM D-4236," indicating that they have been analyzed by a qualified toxicologist.

Public Law 100-695 also forbids schools to purchase any art material labeled for chronic hazards for children in grades below the sixth.

CHOOSING SAFER MATERIALS

Michael McCann, director of the Center for Safety in the Arts, believes that children under twelve should never be allowed to use professional art supplies that contain the hazardous substances discussed earlier in this chapter (metals, solvents, etc.) or whose labels carry special-use instructions or warnings. Children over twelve can be taught proper handling. However, not even older children, McCann feels, should handle materials containing such deleterious substances as asbestos and cadmium.

Other general precautions to take with young children: provide water-based rather than solvent-based products whenever possible, and do not buy dry powders, which can be inhaled, aerosol sprays, or organic solvents. Avoid products with artificial food smells; they encourage children to eat or inhale other materials that may be more dangerous.

For information on safer products for children, call the Center for Safety in the Arts (see Resources). Other helpful sources are two books by Michael McCann, *Artist Beware* and *Health Hazards Manual for Artists*, as well as "Children's Art Hazards" by Lauren Jacobson.

Following are some general guidelines on the use of art materials by children under twelve years. Many of the safer products carry the ACMI's AP, CP, or Health Label Nontoxic seals.

• Avoid artists' pastels, which may contain asbestos-contaminated talc and metal (lead, cadmium) pigments. Also avoid permanent felt-tip pens, which contain toxic organic solvents, and limit the use of water-based markers containing dyes that young children may ingest. Crayons, pencils, dustless chalk, school (dustless) pastels, and poster paints are safer substitutes.

• Children should not work with earth clays that contain silica or asbestos, except when these are wet. Repeated inhalation of clay powder can lead to lung disease. Avoid instant papier-mâché, which may contain asbestos, lead, cadmium, and silica. Super-Dough and Play-Doh are safer. There are also some talc-free clays on the market; AMACO white clay is an example. Papier-mâché can be made from newspaper strips and white paste. An alternative would be an ACMI-certified product like Claycrete (from American Art Clay).

• Don't give your children oil paints that contain toxic pigments. There are water-based as well as nontoxic oil paints on the market (look for the AP/CP seal). However, they should not be cleaned up with toxic organic solvents if children are in the room. A few nontoxic, ACMI-approved cleaners are available, including the Masters Artist Brush Cleaner B & J Original Formula, by General Pencil.

• Professional acrylics may contain metal pigments and should be avoided. Safer substitutes are water colors, school-grade acrylics, and tempera paints. The latter two, however, contain toxic preservatives, so their use by preschoolers should be supervised.

• Don't let your children use professional pottery glazes that contain lead pigment. Leadless glazes are available and labeled as such. Glazes with the AP or CP seal are leadless.

• Avoid airplane, epoxy, and instant-bonding glues, as well as rubber cement and spray adhesives. These contain toxic organic solvents or resins. Safer alternatives are white glue (Elmer's School Glue) or school paste.

• Solvent-based silk-screen inks, and stamping inks, and India inks containing carbon black, a carcinogen—all should be avoided. Use ACMI-certified water-based block print and silk-screen inks instead.

• Replace cold-water fiber-reactive dyes or commercial dyes with vegetable and plant dyes.

TIPS FOR THE SAFE USE OF HAZARDOUS MATERIALS

Below are suggestions for the safe handling of hazardous art materials—if they cannot be eliminated from the household altogether. Of particular importance are the precautions regarding home studios.

• Don't permit eating around art materials.

• Hands should be washed and thorough clean-up procedures followed whenever your children finish using such materials.

• Art studios should be as far away from the kitchen and living spaces as possible.

• Provide good ventilation when working with volatile materials like solvents. This is especially important when a home studio is close to the living area. A good exhaust hood may be a wise investment if a parent is doing acid etching, spray painting, welding, or any other work that releases toxic vapors or dusts into the air.

• All dangerous materials should be clearly labeled, stored in unbreakable, tightly sealed containers, and placed out of reach of children.

• After activities like welding and soldering, all contaminated surfaces, including work benches and floors, should be wet-mopped to prevent the dispersal of dust.

• Hazardous materials required at any stage of your child's art project should be handled by you in the child's absence. When you mix clay powders with water, spray-fix drawings, use aerosols, and clean up with solvents, send your child to another room or outside to play.

If you don't understand a warning label on an art product, or you want more detailed information on the danger of using it, write or call the manufacturer and request "material-safety data sheets." These are forms required by the Occupational Safety and Health Administration (OSHA) for hazardous materials. They contain detailed information about the product, its potential toxicity, and precautions for handling. If the manufacturer won't send these forms to you, and you have doubts about the product's safety, don't buy it.

RESOURCES

Sources of Information

Art Hazards Information Center, Center for Safety in the Arts, 5 Beekman St., Suite 1030, New York, NY 10038. Telephone: (212) 227-6220.

   Publishes an informative newsletter, *Art Hazards News*, and can answer inquiries about art hazards.

Artist Beware. Michael McCann, New York: Watson and Guptill, 1979.

Art and Craft Materials Institute, 715 Boylston St., Boston, MA 02116. Telephone: (617) 266-6800.

   Send your name and address and $2 for its list of certified products.

Artist's Complete Health and Safety Guide. Monona Rossol. New York: Allworth Press, 1991. (Distributed by North Light Books. Telephone: (800) 289-0963.)

"Children's Art Hazards." Lauren Jacobson. New York: NRDC, 1984.

   An informative booklet available from NRDC, 40 West 20th St., New York, NY 10011. Telephone: (212) 727-2700.

Consumer Product Safety Commission, 5401 Westbard Ave., Rm. 318, Bethesda,
   MD 20207. Telephone: (301) 492-6800.
   You can call the CPSC and ask about the safety of specific art materials.
Health Hazards Manual for Artists. Michael McCann. New York: Nick Lyons Books,
   1985.
San Francisco Poison Control Center. Telephone: (800) 233-3360.
   Has a staff toxicologist as well as an extensive data base that can provide
   health information about many toxic substances.

Part 2

The Healthy Yard

Many parents work hard to give their kids a healthy and pleasant outdoor setting in which to play, whether it is the city park, the suburban yard, or the countryside. The last thing parents need is an environment rife with hazards. More and more, that, unfortunately, is what they have to be on the lookout for. Pesticides in the yard, polluted air, increasing levels of ultraviolet (UV) radiation in sunlight—these are just a few. Another concern for parents is how to best protect their children from the annoyance, pain, and sometimes serious illness caused by tick and mosquito bites. The chapters in this section will explore alternatives to pesticides and present safe defensive strategies to use against polluted air, the sun's radiation, and insects.

CHAPTER 5

Maintaining a Chemical-Free Yard

The Problem. In the quest for the perfect-looking yard, suburbanites have become more and more dependent on synthetic pesticides, the majority of which have not been thoroughly evaluated by the EPA for long-term health effects like cancer.

The Risk to Kids. Children love rolling around in the yard, climbing trees, and hiding in bushes, activities that expose them to recently sprayed pesticides. Relative to body size, children take in more of these chemicals than adults and retain them longer, exposing their immature organs to potentially irreparable damage.

What to Do. Parents can avoid using synthetic pesticides, which are unnecessary and in the long run detrimental to the health of your yard because they kill off organisms that control pests. Proper watering, fertilizing, and mowing, as well as the introduction of pest-controlling insects, will keep your lawn healthy. If a serious pest problem develops, there are many safer controls available to the home gardener.

The desire for a beautiful yard has become a suburban obsession in America. Ideal lawns are deep green and blemish free—no weeds, no brown spots, no telltale signs of any living thing except grass. Shrubs, trees, and gardens must appear healthy and vigorous. To achieve this look, suburbanites have put their faith in pesticides—so much so that the volume used on suburban lawns is ten times more per acre than that used on agricultural lands. Pesticide manufacturers and lawn care companies have lured customers with the promise of products that will easily, safely, and effectively eradicate pests and weeds. After pouring chemicals on, your yard may look good, at least temporarily, but at what price to your children? They, after all, are the ones who will be rolling around on the grass, climbing through the bushes and up the trees.

The downside of synthetic pesticide use is suggested in the story of Kevin Ryan of Arlington Heights, Illinois, as told to a U.S. Senate subcommittee investigating lawn chemicals. Kevin can't play outdoors in spring and summer because is he is hypersensitive to the pesticides applied on neighboring lawns. His symptoms include restricted breathing, nausea, memory loss, insomnia, diarrhea, and depression. His mother thinks Kevin's condition began when Kevin was a toddler playing in his sandbox next to a property regularly treated by a commercial lawn company.

Kevin Ryan is a chemically sensitive individual (see below in "Children's Special Vulnerability"). However, there is growing concern over the health effects of pesticides in normal, healthy individuals exposed to low levels over a long period.

WHY ARE CHILDREN VULNERABLE?

Currently, some 600 basic pesticides are marketed in 45,000 to 50,000 different commercial formulations. *Pesticide* is a broad term referring to all pest-killing chemicals, including insecticides, which kill insects, fungicides, which kill fungi and molds, herbicides, which kill weeds and other undesirable plants, rodenticides, which kill rodents, and nematicides, which kill worms.

By definition, pesticides are toxic. They are devised to kill forms of life that are economically, medically, or aesthetically detrimental to man. Since all living creatures have the same basic genetic structure, we can assume that, given the right conditions, pesticides will harm humans as well.

Just how toxic are pesticides for the average child? They have a wide range of effects that depend on concentration, length of exposure, route of exposure (oral, epidermal, etc.), and individual susceptibilities. Acute reactions include rashes, vomiting, and death; chronic effects include cancer and damage to the reproductive tract and nervous system. Although acute effects are documented mostly in workers involved in the manufacture of pesticides or in farmers, they are also seen in children.

Sixty percent of the pesticide cases reported to poison control centers in 1988 involved children under the age of six. Pesticides are the second most frequent cause of poisoning in young children (medicine is the most frequent). Early symptoms of pesticide poisoning—headaches, nausea, eye and skin irritation, and so on—are often overlooked because they mimic other illnesses.

Chronic effects are the most difficult to link with certainty to a particular pesticide. Most studies showing an increased risk of cancer involve farm workers. A 1987 study in *The Journal of the National Cancer Institute*, discovered the risk of leukemia to be up to seven times higher in children under ten whose parents used pesticides in the home or garden at least once a month. The risk was greater for more frequent use.

Another potential chronic effect of these compounds is the development of multiple chemical sensitivity (MCS). Kevin Ryan suffers from this condition, also known as environmental illness, ecological illness, or twentieth-century illness. The cause can be either long- or short-term exposure to a chemical, after which an individual becomes hyperreactive to that chemical as well as to many others. Symptoms, which may involve more than one organ or system simultaneously, include fatigue, headaches, upper respiratory irritation, intestinal disturbance, irritability, and depression. These symptoms may recur with exposure to lower and lower levels of a chemical, making the person progressively less able to function normally.

In the extreme, some MCS sufferers are forced to live in stripped-down rooms to avoid any contact with chemicals. The medical community is divided over whether this is a real illness or an outcome of psychiatric problems. In March 1991, growing media attention and pressure from patient groups led to an EPA-sponsored workshop on developing a research plan for MCS (see chapter 13, "Reducing Harmful Gases").

Children's Special Vulnerability

Studies in animals have indicated that the young are physiologically more vulnerable to pesticides than adults. The reason appears to be the immaturity of the detoxification system in the liver of young animals. The young are also more vulnerable to pesticide vapors because their breathing rate is faster than that of adults. Pesticides may have greater access to the brain in the young. An infant is particularly susceptible to neurotoxic chemicals, that is, those that can damage the brain or nervous system. Neurotoxins can have long-term effects on memory, learning ability, and behavior. As mentioned before, children are more vulnerable to carcinogens because of their young age which allows for the long lag time needed for cancers to develop.

IS THE GOVERNMENT PROTECTING US?

Under the Federal Insecticide, Fungicide, and Rodenticide Act (FIFRA) of 1972, the EPA is required to reevaluate the health and environmental effects of all pesticides registered for use. Many pesticides were registered years ago when there was little knowledge and when testing methods were unsophisticated. As of 1988, the EPA had reevaluated only a handful of pesticide chemicals. To speed up the slow process, amendments to FIFRA in 1988 imposed a time frame of nine years for the EPA to reevaluate these older pesticides in light of current knowledge and to decide whether to reregister them for use. Of the thirty-four most widely used lawn care pesticides, thirty-two are older and subject to reregistration.

So far the EPA has cleared only a few lawn pesticides for reregistration. The public could be at risk until the work is done because all continue to be used. Six of these pesticides—diazinon, 2,4-D, dichlorvos (DDVP), maneb (EBDC), benomyl, and pronamide—have been designated for special review by the EPA because of concerns that they may cause cancer and birth defects, as well as being toxic for wildlife.

Diazinon, the most-used lawn pesticide, has caused massive poisoning of birds. Because of this the EPA has canceled its use for golf courses, but not yet for residential lawns. 2,4-D, the second most widely used lawn care pesticide, is under review by the EPA because of evidence of increased cancer risk to farmers.

Misleading Advertising

The lack of information on the health effects of many widely used yard chemicals calls into question the often bold and misleading advertising done by professional lawn companies that claim that their products are safe. FIFRA authorizes the EPA to take enforcement action against such claims by pesticide manufacturers. The Federal Trade Commission (FTC) also has the authority to protect the consumer against deceptive advertising. Unfortunately, according to a 1990 General Accounting Office study, "Lawn Care Pesticides—Risks Remain Uncertain While Prohibited Safety Claims Continue," the EPA and the FTC have made only limited use of their authority.

HOW ARE CHILDREN EXPOSED?

Recently sprayed pesticides can evaporate in hot weather and be inhaled. Many pesticides penetrate skin, as well as clothing. This means that walking across a freshly sprayed lawn or garden could be hazardous to your child's health.

Homeowners who apply pesticides themselves can inadvertently expose themselves and their children to high concentrations of pesticides by failing to read the instructions on pesticide containers. Pesticides generally come in concentrated form and should be diluted before use. Some people think that if they dilute a pesticide less, it will work better. Others mix and apply pesticides with hands, arms, and legs exposed. If children are hanging around when concentrated pesticides are prepared and applied, they can be exposed to levels that even manufacturers would consider unsafe.

WHAT YOU CAN DO

By now you are probably convinced that it is wise to limit your family's exposure to synthetic pesticides. However, being a busy parent, you don't want to spend long hours toiling in your yard. Here are some simple steps to an attractive and healthy lawn that don't require dangerous synthetic chemicals:

• Change your attitude. Keeping every weed off your lawn or pest off your bushes is next to impossible, even with synthetic pesticides. Be prepared to accept a yard where weeds and insects are present but not out of control.

• Attract and keep natural predators of the pests in your yard. Natural predators include birds, ladybugs, green lacewings, and praying mantises, which feed on pests. You can attract these predators by planting a variety of vegetation such as trees and flowering shrubs. Beneficial insects need pollen and nectar as a source of protein for egg-laying—try planting daisies, sunflowers, wild Queen Anne's lace, and herbs like dill, fennel, and coriander. Mail order companies (see Resources) sell mixtures of herb and flower seeds that attract beneficial insects. These can be planted in your vegetable garden or on the border of your lawn, a project your children may enjoy helping you with. Remember, the best way to keep beneficial insects on your lawn is not to use synthetic pesticides, which kill off the good guys as well as the bad.

• Use repellent plants to keep undesirable insects out of your yard. These can be put alongside the plants pests feed on. Pests, disliking the odor and taste (or some other characteristic) of repellent plants, pick up and move on. Repellent plants include the onion family—garlic, onions, chives, leeks. You will enjoy having them to eat. Herbs such as tansy, borage, spearmint, peppermint, and catnip are also good allies for the gardener. Repellent flowers are helpful in vegetable gardens. Marigolds keep thrips, tomato hornworms, and whiteflies at bay. Geraniums scare away leaf hoppers, and nasturtiums ward off whiteflies and Colorado potato beetles. Consult the books listed in Resources for an extensive list of repellent plants.

• Plant pest-resistant turf grasses suitable to your climate and topography. Many new varieties are available to handle a range of needs: drought-resistant, pest-resistant, cold- and heat-resistant grasses, as well as those that thrive in shade or survive heavy foot traffic (see Resources). For advice on choosing, contact the county extension office of the land-grant university in your state. If you have a backyard garden, you can grow varieties of crops like cucumbers, lettuce, and potatoes that are resistant to common insect pests and diseases. Consult organic gardening books for suggestions (see Resources).

• Build healthy soil. The term "You are what you eat" applies to turf grasses, bushes, trees, and other vegetation in your yard. Soil provides the nutrients and conditions essential for healthy plant growth. Ideally, you should have 5 to 6 inches of rich, dark topsoil. If you don't, add organic matter in the form of topsoil, peat moss, or homemade compost (see composting below).

It is also important to test your soil for acid content. If soil has a low pH, meaning too much acid, lime should be added. If it has high pH, your soil is alkaline and sulfur should be added. Your county extension office is a good source of information, as are local garden centers.

Soil should be free of contamination from deteriorating lead-based house paint and airborne lead emissions from car exhaust, incinerators, or smelters. Lead (discussed in chapter 1, "Limiting Lead Paint and Dust") is a potent childhood neurotoxin. Exterior paint can contaminate the first 10 inches of soil in an area 1 or 2 feet from the house. Always collect exterior scrapings from lead paint with tarps. Airborne lead emissions can settle throughout a yard but tend to concentrate in the area nearest a busy roadside. If your children's play area is contaminated (get advice on testing from your local county agricultural extension), cover the area with about 8 inches of new soil or remove the contaminated soil and add clean soil. Locate your garden away from roads, keep the soil pH-neutral to discourage lead absorption through roots, and plant vegetables like tomatoes (lead concentrates in the leaves and roots, not the fruit, of plants).

• Use organic instead of chemical (inorganic) fertilizer. Organic fertilizer such as well-decayed manure, seaweed, sewage sludge, and compost is made from natural substances and provide water-insoluble, slow-releasing nitrogen as well as important nutrients like phosphorus and potassium. One to three applications during the growing season give a steady supply of nutrients. Commercial products such as Maxicrop, a seaweed product, and Ringer's Lawn Restore (see Resources) are lower in nitrogen content than chemical fertilizers, so they cost more in the short run. However, they save you money in the long run because of the fewer number of applications needed. Unlike many chemical fertilizers, organic fertilizers do not add acid to your soil. Acid is detrimental to the earthworms whose constant movement through soil provide the natural aeration needed for good growth.

Organic fertilizers also prevent your lawn from getting too much nitrogen, a problem with chemical fertilizers. Too much nitrogen makes the grass plant lazy. Its roots no longer have to dig deeply to find nourishment. The resulting shallow roots can contribute to compacting of the soil, a condition which provides a breeding ground for pests that cause turf disease.

• Don't overwater your lawn. Too much watering, especially frequent, light watering, will encourage shallow roots, which during a drought cannot reach water lying deep down. Soil should be wetted to the full depth of the grass's roots—usually 6 to 18 inches deep—and should be allowed to dry out between waterings. It is helpful to dig at least a foot to see how deep the roots are. This will also tell you how long it takes for your soil to dry out after rain or watering. *The Chemical-Free Lawn* by Warren Schultz (see Resources) has a good discussion of watering for different soil types.

• Mow your lawn high and as little as possible. Mowing removes leaf mass, which is critical for photosynthesis. This reduces the plant's ability to produce its own food. It is best to take off no more than four-tenths of the height of the grass at each mowing. Since grass height determines root depth, cutting much closer than this will encourage shallow roots. Higher grass is better because it blocks out sun from emerging weeds, thereby discouraging their growth. Keep your lawnmower blades sharp. Dull blades shred and tear grass, leaving entry points for disease. Leave clippings on your lawn. They break down rapidly and are a good source of nitrogen.

• Keep thatch under control. This is the strawlike, matted layer at the surface of the soil that gives the springy feel to your lawn when you walk on it. Thatch is not, as is commonly believed, the buildup of grass clippings left on the lawn after mowing. Thatch is composed of the tough, fibrous part of the grass plant, which does not decompose unless the right microorganisms and earthworms are present. Too much thatch (more than a half inch) provides the perfect breeding ground for certain turf diseases like dollar spot and fusarium. You can remove thatch from your lawn with a strong wire rake or, for big jobs, with a power rake. Core aeration, which removes small plugs of soil, breaks up the thatch layer and aerates the soil. Also, avoid chemical fertilizer. Heavy doses of nitrogen will kill the bacteria and earthworms that decompose thatch. To help fix a serious thatch problem, add a topdressing of unsterilized topsoil, which contains thatch-busting microorganisms. You can also try Ringer's Lawn Restore, which contains beneficial microorganisms as well as organic fertilizer and is available at garden centers.

• Recycle your yard waste. Because most states are running out of landfill space, yard waste is no longer accepted. Using yard debris (leaves, branches, grass clippings, dead plants) as mulch to put around bushes and on plant beds will help solve the waste problem, prevent weeds, keep soil below from becoming too hot or cold, and add nutrients to the soil.

Composting can recycle large amounts of yard waste and provide you with rich humus for your soil. To build a compost pile, you need some type of enclosure in an unused corner of your yard. Simply add layer upon layer of leaves, grass clippings, weeds, frost-killed flowers, and other organic litter from your lawn. Decomposition can be speeded up by adding unsterilized soil between layers of yard waste or a commercial compost activator, of which there are many on the market. If the pile is working properly, the temperature should reach about 140 to 160°F in four to five days. Five to six weeks later, fork the material into a new pile, bringing the outside of the heap into the center. Add water when the center of the heap dries out. A pile started in the spring should be ready for use by fall. Bacteria, fungi, worms, and other beneficial organisms break the pile down into rich, black humus. Humus loosens heavy clay soil, allows roots to penetrate farther, improves the water-holding capacity of sandy soils, and adds nutrients to the soil. Your local or state department of natural resources can provide you with free information on recycling and composting.

If you follow these steps to a healthy yard, weeds and insect damage should be minimal. In case a yard problem does develop, there are safe, nontoxic ways of treating it. "Integrated pest management" (IPM) strives to reduce pests to manageable levels through physical, mechanical, and bio- logical controls. Chemicals—the most selective—are used only when other methods fail. Specific information on IPM can be obtained from the Bio- Integral Resource Center, a nonprofit organization (see Resources).

The real payoff of natural lawn care is that your yard will be a healthy place for your children to play. An added benefit is that your children will learn a lot about how nature functions if they work alongside you.

WHAT TO DO WITH WEEDS

If you have more weeds than you can tolerate, there are some easy ways to eradicate them. First, get yourself a weed popping instrument, sold at garden centers, and pop the offenders out. You can fill in the resulting bare spots with grass seed or sod. This is an enjoyable activity for young children (they can pull weeds by hand) and good exercise as well. Another way to get rid of weeds is to spray them with the herbicidal soap Sharpshooter (see Resources). It will kill grass and other plants as well, so apply it carefully.

WHAT TO DO WITH INSECTS

The most troublesome pests for the gardener are insects. They can harm plants by chewing or sucking on leaves, stems, and roots and transmitting disease-causing microbes. The adult insect—for example, the fly, butterfly, and beetle—is a nuisance, but the real nemesis of the gardener is the larva (the grub, maggot, worm, or caterpillar), which eats vegetation.

Know Your Enemy
The first step in managing insects is learning which you have, when they appear, and how many there are. Walk in your yard and take a good look at the vegetation. Books on common pests, for example, *Rodale's Garden Insect, Disease, and Weed Identification Guide* and Ringer's *Attack: Organic Pest Control Handbook* (see Resources), will help you put a name to the bugs you see.

Physical Controls
After identifying the invaders, place insect traps (see Resources) around your lawn to get some idea of their numbers. Traps employ colors, shapes, or pheromones (odorous insect chemicals) to lure the prey. Many have a sticky, immobilizing surface. In addition to giving you a good idea of the size of the pest population, traps may be enough to reduce it. An effective physical barrier for crawling insects like tent caterpillars, gypsy moths, and ants is Tanglefoot (see Resources), which can be applied in a band around a tree trunk to prevent insects from climbing up to foliage, buds, and fruit.

Some insects—the Japanese beetle, cucumber beetle, and tomato hornworm, for instance—can be handpicked off vegetation or vacuumed up with a dustbuster. Even your kids can help.

Floating row covers will prevent adult insects from laying their eggs on vegetable plants. A lightweight fabric woven like cheesecloth is placed loosely over plants or seedbeds. Light and water reach the growing plant but insects do not.

Biological Controls

These natural controls—insects, bacteria, and wormlike nematodes—are nature's second best gift to the gardener after the garden itself. Their beauty lies in their ability to target a specific population: they are lethal to a particular type of pest while being harmless to the applicator.

Insects that repress pests can be enlisted in droves to fight on your behalf. All you do is contact a commercial supplier (see Resources) and order insect adults or eggs; they will be shipped to you. The supplier will advise you about which beneficial insects to use for which pests and how to introduce them into your yard. The beneficials most popular with organic gardeners are ladybugs, praying mantises, green lacewings, and trichogramma wasps. You may have to release them several times during the growing season to control a pest.

Milky spore disease is a bacterial infection that kills destructive Japanese beetle grubs and other white grubs that feed on grass roots. A commercial preparation of *Bacillus popillae* spores (see Resources) can be sprayed onto your lawn. Beetle grubs will ingest the spores and die. This treatment is not a quick fix; it may take a season or two to bring the pests under control. However, the bacterial spores will remain dormant in your soil, ready to be consumed by beetles that show up even years later.

Nematodes are a safe weapon for your arsenal against pests that crawl in the soil—Japanese beetle grubs, sod webworms, strawberry root weevils, and cabbage maggots. Nematodes are wormlike creatures, but don't worry—they are so small you can't see them. They enter the bodies of their prey and release deadly bacteria. Nematodes are shipped in a topsoil or peatmoss-like medium (see Resources) that is watered into the soil.

Bacillus thuringensis (B.t.) produces a toxin that kills caterpillar larvae within two days of ingestion. It is active against hundreds of caterpillar species, including armyworms, sod webworms, codling moth larvae, gypsy moth larvae, tent caterpillars, and cabbage loopers. B.t., which comes in powder form, is dissolved in water and sprayed onto plants (see Resources).

Chemical Controls

Natural-occurring chemical insecticides are safer than synthetic pesticides because of their low toxicity and biodegradability. They can be used for a serious pest problem when the strategies discussed above aren't working.

Insecticidal soaps are made of potassium salts of fatty acids derived from animal fat, milk, and plant oils. They disrupt the external membrane of an insect, causing it to lose body fluid and eventually die. Insecticidal soaps can be sprayed on plants, fruits and vegetables, and turf. They are generally used for killing soft-bodied insects like aphids, mites, scales, whiteflies, mealybugs, and chinch bugs. Safer and Ringer, which sell a variety of formulations for your particular pest problem (see Resources), are available at most garden stores.

Neem oil, extracted from the Neem tree, has been used as an insecticide in India for centuries without apparent harm to humans. It makes plants unpalatable to hundreds of insects, including leafminers, aphids, chinch bugs, gypsy moths, and thrips.

Pyrethrin, extracted from chrysanthemums, has been used as an insecticide for over two thousand years. It penetrates the outer cuticle of insects, rapidly poisoning the nervous system. Pyrethrin-based products (see Resources) are effective against large, hard-bodied insects like Japanese beetles, Colorado potato beetles, and cucumber beetles, as well as sawfly larvae. Apply the product to limited areas; being nonspecific, it will also kill beneficial organisms.

Pyrethrin can produce skin irritation, although this is rare. Some commercial products contain a small amount of piperonyl butoxide (PBO), which may be toxic to the liver. You should wear gloves and good coveralls when applying products containing PBO. Pyrethroids are synthetic chemicals similar in structure to pyrethrin but much more toxic and less biodegradable—they should be avoided.

WHAT TO DO WITH VIRUSES, BACTERIA, AND FUNGI

Unlike insect pests, microorganisms that infect plants are often hard to recognize. They cause turf diseases like fusarium blight, and dollar spot, gray mold on grapes and downy mildew on lettuce. Viruses, bacteria, and fungi may be spread to plants by insects. Birds, animals, and gardeners can also carry disease-causing microbes from plant to plant.

A well-balanced soil containing the right amount of nutrients and beneficial organisms will keep plants healthy and better able to resist attack by disease-causing microorganisms. Follow the steps to a healthy yard mentioned earlier, remembering especially to limit broad-spectrum pesticides that will also kill natural predators of the pests that transmit diseases to your plants. You can avoid spreading plant disease by cutting off and destroying diseased vegetation, and washing your hands before touching other plants. If you have a garden, rotating crops is a good way to keep soil, and therefore crops, healthy.

IF YOUR CHOICE IS PROFESSIONAL LAWN CARE

If you don't have the time to take care of your lawn and rely instead on a professional lawn care company or landscape service, choose carefully. More and more such companies are adding less toxic methods to their list of services. Call around and ask companies if they will use IPM on your lawn. If companies in your area only use synthetic pesticides, tell them you would prefer spot treatment to blanket spraying. Find out what pesticides are used and ask for material safety data sheets on each one. These sheets provide detailed information on the health effects of particular pesticides as well as emergency and first-aid procedures to take following exposure.

RESOURCES

Products and Services

Grace-Sierra. Telephone: (215) 395-7104.
    Call for information on the Neem oil product Margosan-O.
Lofts Seed, P.O. Box 146, Bound Brook, NJ 08805. Telephone: (800) 526-3890.
    Sells pest-resistant grasses.
Maxicrop, P.O. Box 964, Arlington Heights, IL 60006. Telephone: (708) 253-0756.
    Markets Maxicrop, an organic, seaweed-based fertilizer.
Natural Gardening Research Center, Highway 48, P.O. Box 149, Sunman, IN 47041. Telephone: (812) 623-3800.
    Sells a variety of less toxic yard care products, including flowering herbs to attract beneficial insects, sticky insect traps, Tanglefoot, floating row covers, and beneficial insects. Write for a free catalogue.
Necessary Trading, P.O. Box 305, New Castle, VA 24127. Telephone: (800) 447-5354.
    Puts out a catalogue from which you can order B.t., nematodes, milky spore disease, composting products, and organic fertilizers.
Peaceful Valley Farm Supply, P.O. Box 2209, Grass Valley, CA 95945. Telephone: (916) 272-GROW.
    Sells a variety of less toxic products, among them beneficial insect attractants and pyrethrin insecticides. Also offers a soil analysis service and advice on composting. Call for a free catalogue.
Rincon-Vitova Insectaries, P.O. Box 95, Oak View, CA 93022. Telephone: (800) 248-BUGS.

Sells a wide variety of beneficial insects.

Ringer, 9959 Valley View Rd., Eden Prairie, MN 55344. Telephone: (800) 654-1047.

Sells many less toxic lawn care products, including Lawn Restore organic fertilizer, milky spore disease, B.t., and pyrethrin insecticides. Request a copy of *Attack: Organic Pest Control Handbook*, an illustrated guide to common garden pests and organic control methods.

Safer, 189 Wells Ave., Newton, MA 02159. Telephone: (800) 423-7544.

Markets a variety of herbicidal and insecticidal soaps.

Turf Seed, P.O. Box 250, Hubbard, OR 97032. Telephone: (503) 981-9571.

Sells pest-resistant grasses.

Sources of Information

The Bio-Integral Resource Center, P.O. Box 7414, Berkeley, CA 94707. Telephone: (415) 524-2567.

Publishes informative pamphlets, including "What Is IPM?" and "Least-Toxic Pest Management: Lawns."

The Chemical-Free Lawn. Warren Schultz. Emmaus, PA: Rodale Press, 1989.

Provides information on pest-resistant grasses as well as nontoxic lawn care principles and practices.

Citizens for a Better Environment, 33 E. Congress St., Suite 523, Chicago, IL 60605. Telephone: (312) 939-1530.

Prints fact sheets on a variety of lawn care issues like composting and pesticide use.

The Encyclopedia of Natural Insect and Disease Control. Roger B. Yepsen, ed. Emmaus PA: Rodale Press, 1984.

A good source of information on pest-resistant and repellent plants.

National Coalition against the Misuse of Pesticides, 701 E St. SE, Suite 200, Washington, D.C. 20003. Telephone: (202) 543-5450.

Write for the brochure "Least Toxic Control of Lawn Pests" (50¢) as well as a publications list.

Rachel Carson Council, 8940 Jones Mill Rd., Chevy Chase, MD 20815. Telephone: (301) 652-1877.

Has a brochure, "Pesticides in Contract Lawn Maintenance," which lists commonly used synthetic pesticides and their health effects, and a publications list.

Rodale's Garden Insect, Disease, and Weed Identification Guide. Miranda Smith and Anna Carr. Emmaus, PA: Rodale Press, 1988.

University of Nebraska Cooperative Extension Service, Bulletin Room 105 ACB, Lincoln, NE 68583-0918. Telephone: (402) 472-3023.

Write for the following illustrated booklets ($3 each): "Turfgrass Disease Damage Prevention and Control" (EC 81-1235), "Turfgrass Insect Damage Prevention and Control" (EC 81-1238), and "Turfgrass Weed Identification and Control" (EC 83-1241).

CHAPTER 6

Keeping Biting Pests at Bay

The Problem. Children playing outdoors are exposed to ticks and mosquitoes that can transmit serious illnesses like Lyme disease and encephalitis. Parents generally rely on commercial insect repellents to protect their kids. However, these repellents are not safe for application to young children's skin.

The Risk to Kids. The health effects of repellents range from skin irritation to, in rare cases, toxic encephalopathy, a brain disorder that can result in convulsions and death.

What to Do. Protect your children with as much clothing as possible, and apply repellent to the clothing, never the skin. Try to reduce the mosquito and tick populations in your yard by utilizing integrated pest management (IPM).

In the summer of 1975, eleven-year-old Todd Murray of Lyme, Connecticut, began having symptoms of arthritis. His doctor later diagnosed his condition as juvenile rheumatoid arthritis (JRA). Interestingly, his mother began hearing of other children in the area who were also diagnosed with the disease. Realizing that the cluster of sick children, some forty in total, was unusual, Mrs. Murray notified the state health department. Meanwhile, another mother, in nearby Old Lyme, identified another cluster of JRA cases, including her own daughter. The mothers' astute observations marked the beginning of the unraveling of the mystery of Lyme disease. In time, scientists would link the disease, of which arthritis is one symptom, with tick bites.

TICKS AND YOUR CHILDREN'S HEALTH

In addition to transmitting Lyme disease, ticks can transmit relapsing fever, Rocky Mountain spotted fever, and Colorado tick fever.

Ticks, the close relatives of spiders and scorpions, are primarily parasites of wild animals. However, as people encroach more and more on once virgin forest, they and their domestic animals are becoming targets as well. Ticks attach to the host (animal or human) and suck blood, often without being noticed. After having a meal they can transmit disease-causing organisms to new hosts.

Your family pet, especially if he runs free in the woods, may be a source of exposure to ticks for your children. Children also encounter ticks while hiking, camping, or playing in a yard either near a wooded area or frequented by deer and rodents (hosts for some ticks). Ticks are generally not a problem in urban areas.

The primary disease-causing tick, the so-called hard tick, can be recognized from the top by its tapered head, visible mouthparts, and the shieldlike plate on its back. Different species can transmit the same disease. For example, Rocky Mountain spotted fever is transmitted by the American dog tick in the East and by the Lone Star tick in the Southwest. Lyme disease can be transmitted by the northern deer tick in northeastern and midwestern states and by the western black-legged tick in western states.

Lyme Disease

Lyme disease is currently one of the most feared of the illnesses transmitted by ticks. It is caused by a corkscrew-shaped bacterium that ticks pick up from the blood of deer and rodents. Named for the Connecticut town where it was discovered, Lyme disease has now been detected in every state except Alaska, Hawaii, Nebraska, Montana, and Arizona. New York, New Jersey, Connecticut, Massachusetts, Rhode Island, and Pennsylvania account for some 90 percent of all cases. Wisconsin, Minnesota, and Georgia also report a large number of cases. Children playing outdoors in these regions from early spring to early fall (until the end of November in warm climates) are most at risk.

In humans, the disease typically begins as a raised red rash with a "bull's eye" where the tick attached. The rash appears three to ten days following the bite. However, 25 percent of victims never develop a rash. Flulike symptoms of fever, headache, stiff neck, joint pain, and fatigue may occur as well, and may disappear with or without antibiotic treatment. Weeks or months later, more serious symptoms—facial paralysis, arthritislike pain in the joints, and cardiovascular and neurological problems—may occur if the disease is left untreated.

Without a definitive diagnostic test for Lyme disease, doctors often err on the side of caution by prescribing antibiotics. These are generally effective if the disease is caught early enough, but they may have little effect once it progresses. Current research on Lyme disease promises to yield a human vaccine in four to eight years.

Other Diseases Caused by Ticks

Relapsing fever often occurs in individuals who visit recreational areas with rodent-infested campsites and cabins. Five to eleven days after being bitten by an infectious tick, the victim experiences a sudden onset of high fever with chills, headache, and muscular tenderness. Fever lasts about three to six days and recurs, but no more than ten or so times.

Rocky Mountain spotted fever is most common in individuals under fifteen. Epidemics tend to occur in relatively small areas like Long Island (and not, as the name would suggest, the Rocky Mountains). The disease-causing agent is a microorganism called a rickettsia, which ticks pick up from chipmunks, squirrels, raccoons, and other hosts. Fever, headache, rash, mental confusion, and muscular tenderness occur two to eight days after a bite. The rash is the most distinguishing feature of this disease, appearing first on the wrists and ankles and spreading within hours to the trunk. Rocky Mountain spotted fever can affect intellectual functioning in children and in some cases lead to learning disabilities.

Colorado tick fever, caused by a virus, occurs in the Rocky Mountain states and the northern Sierra Mountains of California. Symptoms include fever, headache, backache, and malaise and occur five to ten days after a bite. In children, the disease can produce serious complications of the nervous system.

How You Can Protect Your Children

The following guidelines will help you protect your children from diseases transmitted by ticks:

• Cover your children in clothes when they play in woods or grassy areas. Long pants, long-sleeved shirts, closed shoes, a hat that covers the ears—these are all necessities. Dress children in light-colored clothing (this will help you spot a tick, about the size of a pin head), and tuck their pants inside their socks. Of course, your children will protest—wearing clothes from head to toe on a hot day is no fun. Explain that you are protecting them against ticks; the prospect should be enough to make the most recalcitrant child comply. If not—if your child insists on shorts rather than long pants—then pick the longest shorts you can find and the tallest socks.

• Apply an insect repellent to the clothes, not the skin. Costly tick repellents are not necessary. Repellents don't kill insects, they just discourage them from biting. A variety of repellents are on the market, the most effective of which are those containing deet. It repels more species of bugs and for longer than other chemicals. OFF! and Cutter both contain deet. Deet-containing repellents will also ward off mosquitoes, chiggers, fleas, and biting flies like the black fly, whose bite can be severe.

For small children, it is best to use repellents with lower concentrations of deet. Again, apply only to the clothing. This will prevent possible side effects like rashes or toxic encephalopathy, a potentially severe brain disorder.

Parents should be aware that deet is also a solvent for many plastics such as those in eyeglass lenses or in watch crystals. It can be safely used on cotton, wool, and nylon but can damage synthetic fibers like rayon, acetate, and spandex (in bathing suits).

A new insect repellent containing permethrin, a synthetic pyrethrin, is receiving much attention. It is the only product that was shown to be 100 percent effective against ticks (deet is only 85 to 95 percent effective), and it kills rather than just repels them. Permethrin, sold under the name Permanone Tick Repellent, is currently available in about thirty-one states. The manufacturer does not recommend its use on children or their clothing; however, the Bio-Integral Resource Center, a nonprofit organization that provides information on the least toxic methods for managing pests (see Resources), recommends it be applied to the clothing of children playing in tick-infested areas. The small risk from permethrin is worth the added protection it provides against potentially serious tick bites.

• Check your children every hour or so for ticks. Ticks feed on their hosts anywhere from two to twenty-four hours before transferring disease organisms, so the sooner you spot a tick and remove it, the better your child's chance of avoiding infection. At the end of the day, during bath time, examine your children closely, going over the scalp, ears, navel, and groin. Ticks that have not yet attached to the skin can simply be washed or brushed off.

• To remove an attached tick, use a pair of blunt-ended, curved tweezers (see Resources). Place these around the tick's body and as close to your child's skin as possible. Then gently but firmly pull the tick straight out. Clean the bite with antiseptic, and monitor it for several days. If changes like darkening or inflammation occur, consult your doctor. Preserve the tick in a jar of rubbing alcohol so your physician or county health department can determine what kind it is. If you don't have tweezers, wear thin gloves to remove the tick from your child. You can also try coating the tick with petroleum jelly, which may force it to withdraw within thirty minutes for lack of oxygen. Other methods, such as applying a lighted cigarette or hot needle to the rear of the attached tick, are not recommended by experts.

• If ticks are a menace in your area, control them (and rodents) by using IPM. Mow your grass regularly, and cut down brush, especially around play areas. Keep garbage cans securely closed and firewood and yard waste away from the house and play area. Remove bird feeders that can be reached by rodents. For a serious problem you can try Damminix (see Resources). It consists of permethrin-soaked cotton in small cardboard tubes. The tubes are placed in brush where mice nest. Mice take the cotton back to their nests, where the permethrin kills the ticks but does not appear to harm the animal.

If deer regularly pass through your property, you might consider fencing to keep them out.

• Check your dog or cat regularly for ticks. Household pets may pass ticks on to you and your children. Behind the animal's ears, around the neck, and between the toes are areas where ticks often lodge. A masking-tape lint roller can be run over the body of your pet to capture unattached ticks. Don't let your pet wear a tick/flea collar continuously. Many collars contain toxic pesticides whose vapors can be inhaled by the whole family. Put the collar on your pet when it is in tick-infested areas. Between uses, store the collar in a tightly sealed container. Low-toxicity insecticides composed of citrus oil extracts or pyrethrins (see Resources) are available from pet stores and can be used to treat your pet, its bedding, and other areas it frequents. A cat may be irritated by citrus oil products; test them first in small amounts.

Remove attached ticks from your pet as you would from your child.

MOSQUITOES AND YOUR CHILDREN'S HEALTH

Mosquitoes can be a serious health threat if they carry viruses that cause encephalitis. Epidemics of this disease usually occur somewhere in the United States every summer. In its severest form, encephalitis attacks the central nervous system, resulting in convulsions, muscle weakness, paralysis, or mental retardation, outcomes more frequent in children than in adults.

Encephalitis is usually contracted in wild bird habitats like woodland swamps and forested areas. Wild birds carry the disease-causing virus, which is transmitted to the human via the mosquito.

How You Can Protect Your Children

To protect your children from mosquito bites as well as encephalitis, consider the following tips:

• Spray repellents on your children's clothing. Deet-containing products are the most effective. Alternatives are citronella- and herbal-based formulas (see Resources); these products work, but not as well or for as long as deet. Also, they can be a skin irritant for some people. Since the effects of citronella- and herbal-based repellents are not completely known, they should be used only on clothing, not skin. Avon's Skin So Soft Bath Oil has been used as a popular, word-of-mouth remedy for mosquitoes. An alternative strategy is to eat lots of garlicky food, because mosquitoes hate the smell. Basil repels mosquitoes; placing a few plants around the home may help keep mosquitoes away.

• Try to control the mosquito population around your home. This is the place your children will most likely be bitten. Since the larval stage of all mosquitoes is aquatic, even the smallest container of standing water can be a breeding ground for a large number of mosquitoes. Don't leave water standing in birdbaths, kiddie pools, gutters and spouts, old tires, saucers under potted plants, or animal dishes. Also, don't overwater your lawn.

Small backyard ponds can be stocked with mosquito-eating fish like goldfish and gambusia fish. The bacterium *Bacillus thuringiensis israelensis* (B.t.i.) (see Resources) is a highly selective, rapid-acting stomach poison for mosquito larvae that can be applied to ponds and water-logged lawns. B.t.i. is also effective against many species of black flies.

The insect growth regulator (IGR) methoprene (see Resources) attacks mosquito larvae. IGRs are hormones that control part of the insect's life cycle, such as the transformation of a larva into an adult. Since humans do not undergo such transformation, IGRs are safe to handle. One methoprene product is a slow-release briquet (Altosid, by Zoecon) that can be placed in standing water. This product is sold to communities for mosquito abatement and may possibly be obtained for use in your own pond by contacting your local mosquito abatement agency or pest control company. For further information and advice on these and other, less toxic control strategies, contact the Bio-Integral Resource Center (see Resources).

• Find out whether the agency responsible for mosquito control in your area is using or considering the use of any of the methods discussed above. They are safer for humans and far more selective for mosquitoes than the organophosphate or carbamate insecticides some communities apply by fogging or aerial spraying. If you are concerned about spraying in your area, contact the local chapter of the League of Women Voters. They are involved actively in evaluating the safety of mosquito abatement programs.

RESOURCES

Products and Services

Abbott Laboratories, 1400 Sheridan Rd., N. Chicago, IL 60064. Telephone: (800) 323-9597.

> Sells a B.t.i. product called Vectobac for mosquito control in ponds.

BioQuip, 17803 S. LaSalle Ave., Gardena, CA 90248. Telephone: (213) 324-0620.

> Sells blunt-ended, curved tweezers for tick removal.

Co-op America Order Service, Dept. GM, 49 The Meadows Park, Colchester, VT 05446. Telephone: (202) 223-1881.

> Markets Green Ban, an herbal-based mosquito repellent.

EcoHealth, 33 Mt. Vernon St., Boston, MA 02108. Telephone: (800) 234-8425.

> Sells Damminix for tick control in mice.

Farnam Pet Products, P.O. Box 34820, Phoenix, AZ 85067. Telephone: (800) 343-9911.

> Markets Flea Stop Flea Mist (pyrethrin-based) for controlling ticks and fleas on your pet.

Gardener's Supply, 128 Intervale Rd., Burlington, VT 05401. Telephone: (802) 863-1700.

> Sells Natrapel, a citronella-based mosquito repellent. Call for their catalogue.

Natural Pest Controls, 8864 Little Creek Drive, Orangevale, CA 95662. Telephone: (916) 726-0855.

> Sells mosquito-eating fish.

Necessary Trading Company, P.O. Box 305, New Castle, VA 24127. Telephone: (800) 447-5354.

> Has Citrus Oil Spray and Shampoo to repel ticks and mosquitoes from your pet, and "Shooo" insect repellent, an herbal oil mix for adding to tick/flea collars.

Ringer, 9959 Valley View Rd., Eden Prairie, MN 55344. Telephone: (800) 654-1047.

Sells Flea and Tick Attack, a pyrethrin-based insecticide for treating your pet.

Safer, 189 Wells Ave., Newton, MA 02159. Telephone: (800) 423-7544.

Markets Flea and Tick Spray (pyrethrin-based) for tick control on your pet.

Zoecon, 12005 Ford Rd., Suite 800, Dallas, TX 75234. Telephone: (800) 527-0512.

Markets B.t.i. under the name Teknar, and an insect growth regulator for mosquito control, Altosid.

Sources of Information

The Bio-Integral Resource Center, P.O. Box 7414, Berkeley, CA 94707. Telephone: (415) 524-2567.

Provides information on IPM methods for specific biting pests.

Common-Sense Pest Control. William Olkowski, Sheila Daar, and Helga Olkowski. Newton, CT: Taunton Press, 1991.

Gives least-toxic solutions for control of home and garden pests, including ticks and mosquitoes.

CHAPTER 7

Breathing Defensively Outdoors

The Problem. America's lower atmosphere is polluted by evaporated organic chemicals and fuel-combustion byproducts, many of which are toxic or carcinogenic, and some of which cause problems as diverse as city smog and acid rain.

The Risk to Kids. Children get outside more than adults and are therefore more exposed to outdoor air pollution. Relative to body size, children also breathe in more air than adults and thus absorb more air pollution. As a result, some children experience reduced lung function and lowered resistance to respiratory disease, and some run an increased risk of cancer.

What to Do. Immediate steps include reducing children's exposure to hot-weather ozone in city smog and consulting the Toxics Release Inventory (TRI) to learn what is polluting the air in your community. Parents can also support clean air regulations, and adopt consumer and driving strategies that reduce air pollution.

Clean air is a mixture of nitrogen and oxygen with traces of other gases. Traditionally, dirt and smoke made air dirty, while breezes and rain regularly cleaned it up. But clean air is a scarce commodity today, and not only around obvious places like cities, industrial sites, and major roadways. The massive amount of manmade air pollution produced yearly—over 200 million tons—can envelop large regions of the country at a time and reach right into backyards, parks, and summer camps.

Polluted air contains hundreds of different toxic gases and suspended particles. Does it affect children? Simply put, yes. Researchers at the

National Center for Health Statistics suspect that a steady rise in asthma deaths among children since the late 1970s may be due to increasing air pollution. Healthy children with no chronic respiratory problems are also suffering. Several recent U.S. and Canadian studies show that current levels of toxic ozone and other air pollutants are associated with loss of pulmonary function and increased hospitalization for respiratory disease. One study, done in 1984, examined how air pollution affected children at a YMCA camp in the New Jersey countryside, a spot well away from any major source of pollution. Researchers from New York University measured lung function in the children as they played outdoors in hazy summertime air, which contains ozone and other pollutants. When ozone levels approached (but didn't exceed) the legal limit, most children exhaled less air than normal, a significant sign of reduced lung function. Ozone gas, which irritates lungs and interferes with oxygen uptake, is only one of hundreds of chemical pollutants in the air.

WHAT POLLUTES AIR AND WHERE DOES IT COME FROM?

Air is polluted by tiny fibers and particles, metals, and hundreds of gaseous chemicals that together make up a brew of carcinogens, neurotoxins, and plant and animal poisons. Pollutants come mainly from factory and power plant emissions, motor vehicle exhaust, and evaporated household chemical products. The main culprits have been gradually identified and many of their sources regulated over the last twenty years. Two important pieces of legislation helped achieve this: the 1970 Clean Air Act, and the 1986 Emergency Planning and Community Right-to-Know Act, which established the TRI.

Identifying the Pollutants

The 1970 Clean Air Act fingered six substances as the major contaminants of air: ozone, hydrocarbons, particulates, nitrogen dioxide, sulfur dioxide, and carbon monoxide. The EPA set limits on these pollutants and in 1978 added the metal lead to the list. It turns out that many more chemicals and metals contribute to air pollution, but this first group of regulated pollutants constituted—and in some cases still does—a serious environmental hazard.

At ground level, ozone gas is well known to cause coughing, pain, shortness of breath, and eye irritation. There is growing evidence that it causes permanent lung damage. It is not directly emitted by cars or

industry stacks but rather is formed by a chemical reaction between sunlight and other stagnant air pollutants. During hot, sunny weather, gas
fumes, evaporated organic chemical solvents, carbon monoxide, and nitrogen dioxide mix to form ozone, the main component of yellow-brown urban
smog. (This ozone is not to be confused with "good," stratospheric ozone;
see chapter 8, "Sunning Safely.")

Hydrocarbons are a large family of organic chemicals including fuels and
solvents, examples of which are butane and octane. Hydrocarbons have
many sources. They leak into the air from refineries, chemical plants, dry
cleaners, print shops, and automobile emissions. Many hydrocarbons are
highly toxic, some are well-documented carcinogens, and they contribute
to ground-level ozone. Children are regularly exposed at gas stations to one
of the polluting hydrocarbons, benzene, in concentrations 300 times the
normal outdoor level. Benzene released from filling up cars, car exhaust,
and solvent products is thought to cause 99 percent of the 960 benzene-
induced leukemias per year in the United States.

Particulates are tiny particles suspended in air. They include dirt, soot,
and smoke from factory, car, and power plant emissions. Some of the tiniest
particulates contain toxic metals and chemicals. They are small enough to
penetrate deeply into lung tissue, where they contribute to respiratory
disease and release hazardous material into the bloodstream.

Nitrogen dioxide and sulfur dioxide are both inorganic chemicals that
form from burning fossil fuels like gas and coal. Burning gasoline produces
high amounts of nitrogen dioxide, which contributes to ground-level
ozone. Burning high-sulfur coal produces high amounts of sulfur dioxide.
Both these gases change chemically when they enter the atmosphere,
forming acid aerosol particles. There is growing evidence that they contribute to respiratory illnesses like chronic cough and bronchitis in children.

Carbon monoxide is a colorless, poisonous gas that is a byproduct of
burning auto fuel. In low concentrations, carbon monoxide can produce
fatigue in healthy people. In the high concentrations found in congested
cities, it can cause headache and dizziness. It is a major contributor to
ground-level ozone.

Lead is a toxic metal that enters the air from industrial emissions and
from vehicles burning leaded gas. Inhaled lead particles enter the bloodstream and are toxic to all tissue, especially nerve tissue. Lead emissions
that are not inhaled settle on the ground, contaminating gardens, lawns,
and cropland (see more on this in chapter 5, "Maintaining a Chemical-Free
Yard"). The use of leaded gas from the 1920s through the 1970s has left
huge quantities of lead, which does not biodegrade, in roadside soil.

Unmasking Air Toxics

Although national emission standards were eventually set for a small number of other pollutants, such as asbestos and vinyl chloride, the Clean Air Act of 1970 didn't address the hundreds of other chemicals and metals polluting the air. No one, not even scientists knew what kind of "air toxics" were being released. However, in 1986 Congress passed the Community Right-to-Know Act, which requires industrial facilities to estimate and report for the TRI their annual release of toxic chemicals to air, land, and water. The first TRI revealed that in 1987, industries across the nation released 235 million pounds of carcinogens and 527 million pounds of neurotoxins. The EPA calculated that just 20 of the 329 chemicals in the survey cause more than 2,000 cases of cancer every year.

In a number of towns and states where industrial emissions are high, citizens have formed action groups to press for reduced emissions. Some chemical industries, including Monsanto, Union Carbide, Du Pont, and 3M, pledged voluntarily to reduce the emission of air toxics. And the government's updated Clean Air Act, passed in 1990, specifically limits industrial air toxics emissions. This is welcome action, since the total amount of air pollution still isn't known. There is no account of the emissions from waste dumps, car exhaust, dry cleaners, or gases that evaporate from chemicals released into soil and water.

WHY ARE CHILDREN VULNERABLE?

Several factors make outdoor air pollution a serious hazard for children. Children play outside, especially during the summer when the ground ozone level is high. They inhale thousands of liters of air every twenty-four hours, and their respiratory system absorbs pollutants and releases them into the bloodstream. Children are more vulnerable than adults to air pollution because they breathe at a faster rate, taking in more airborne pollutants, and closer to the ground, where many pollutants concentrate.

Effects on Children

There is growing medical evidence that chronic exposure to air pollution causes reduced lung capacity, bronchitis, and permanent lung damage. A study in Glendora, California, has tracked such a situation. Glendora sits at the foot of the San Gabriel mountains and traps smog moving east from Los Angeles. The lung function of citizens aged 7 to 59 years was compared to that of people in a nearby town protected from Los Angeles's air pollution by a mountain range. Normally, lung capacity increases during the teenage years, and then goes into a steady decline. However, children's lung capacity grew more slowly, and adults' declined more rapidly in Glendora than in the less polluted town.

Doctors are seeing a dramatic increase in asthma. The National Center for Health Statistics reports that asthma cases among children aged six to eleven years increased 58 percent during the 1970s. Between 1979 and 1987, asthma death rates almost doubled for children aged five to fourteen. Today about 7 percent of American children have asthma, most commonly inner city blacks. Researchers suspect that air pollution is contributing to the increase in asthma cases and deaths. For more on this issue, see chapters 13 and 14.

DOES CLEAN AIR LEGISLATION PROTECT KIDS?

Once the 1970 Clean Air Act set federal standards for the six major pollutants, considerable progress was made in reducing auto pollutants. In fact, exhaust from new cars is 96 percent cleaner than exhaust from cars built before 1970, thanks largely to pollution-control devices like the catalytic converter. This device turns emitted hydrocarbons and carbon monoxide into nonirritating carbon dioxide and water. The Clean Air Act also made factories that emit more than 100 tons of smog-forming chemicals a year adopt air pollution equipment. In spite of such progress, standards for ozone and carbon monoxide were not met.

What dealt the first clean air effort a blow was a massive increase in air pollution from an increasing number of polluters. There was also the problem of defining acceptable levels for specific pollutants. Regarding sulfur dioxide and nitrogen oxide, for example, what seemed a safe limit in 1970 proved to be inadequate. It failed to stop the formation of ozone and acid rain. The cap on particulates targeted total suspended particulates (TSPs) and measured them by weight. Although TSP levels were brought down to the federal limit, the light, fine particulates—which, being breathed in most deeply, are the most hazardous—escaped regulation. In 1987, the EPA reduced the size of regulated particulates. But recent research indicates that not even the new standard is sufficient to protect children's respiratory health. A study of a community exposed to air pollution from a Utah steel mill found a correlation between hospital admissions for children's respiratory illness and fine particulate levels that did not exceed the new government standard.

The Latest Battle Plan

America has high expectations for the 1990 amendments to the Clean Air Act. The amendments are 868 pages—the 1970 act was only 41—and in these pages are measures environmentalists hope will reduce the thousands of deaths that occur annually because of air pollution. The new act has three main goals: to bring the ozone in city smog under control, to stop acid rain, and to reduce air toxics.

New controls are planned for cars and industries that emit substances that create ozone. Fuels themselves will be reformulated to evaporate less and burn more cleanly. Fumes that leak at gas stations during fill-ups not only pollute the air but also expose children waiting in cars to an unacceptable concentration of hazardous hydrocarbons. New cars will feature fume-catching canisters and gas tanks will have "vapor-recovery nozzles." Auto makers are to begin producing cars that run on cleaner fuels like natural gas or methanol. Among other measures mandated by the 1990 amendments are new controls on air toxics. Industries are required to reduce by 90 percent the emissions of 189 toxic chemicals, although they have until the year 2000 to do this. And to reduce acid rain, power plants and factories will cut sulfur dioxide emissions roughly in half from 1980 levels by the year 2000, and cut nitrogen oxides about 10 percent a year beginning in 1995.

There are several problems with the new clean air legislation is it relates to children. First, it ignores their special vulnerability. U.S. health standards are based on what the healthy adult male can tolerate. Second, the legislation fails to consider the possible hazard created by the interaction of pollutants. Controversy exists about what constitutes safe levels of ozone and acid particles. And third, although the tough new measures promise to protect children's respiratory health, they won't have a measurable effect until after the turn of the century. In the meantime, parents should take special steps to protect their children's lungs.

GIVING YOUR CHILDREN CLEAN AIR, NOW

Unfortunately, you can't go out and buy a yardful of clean air. But there are two things you can do in the short term. The first is to find out what air toxics there might be in your area by consulting the TRI. You may want to consider air quality when deciding where to live and raise your family. The second is to listen to air-quality reports before your children go outside to play. Perhaps they should not be outside exercising vigorously on summer days when the pollution is high.

The Toxics Release Inventory

The EPA has made the TRI available on microfiche to libraries across the nation. You can look up the names of industries in your area and find out what pollutants they are releasing to the atmosphere. The EPA also provides fact sheets describing the health effects of different pollutants. If you have a pressing or specific question, call the EPA's Emergency Planning and Community Right-to-Know Hotline (see Resources). The TRI is also available on computer disk from the National Library of Medicine's TOXNET database (see Resources). You can search out a company by name and then read the list of substances it releases in your area. Or you can search out a particular chemical to find how many pounds have been released in your zip code, county, city, or state. The TRI will also give you information on how many pounds of a chemical have been shipped to your state for disposal. One other source of information on emissions is *A Who's Who of American Toxic Air Polluters*, published by the Natural Resources Defense Council (see Resources).

What is the good of finding out about pollutants in your air? Consider what happened when citizens of Northfield, Minnesota, discovered that a local industry was the forty-fifth largest emitter of methylene chloride, a probable carcinogen, in the nation. Community groups organized to get the industry to clean up its emissions. A petition was circulated among people living closest to the plant, and faculty at local colleges reviewed the health threat of the contaminant. Legislators were asked to use their influence to get the state pollution control agency to delay issuing the plant a new permit until the emissions issue could be explored. Both the state agency and the plant itself developed plans to reduce the emissions. The point of this story is that citizens can refuse to accept the risks imposed on them by a local industry's air toxics. Citizens can demand better monitoring of emissions and tougher state laws to control toxic releases.

Breathing Defensively

You can help protect your children from the ozone, carbon monoxide, and acid particles in summertime haze by following the rules proposed to protect joggers and the elderly.

First, familiarize yourself with the smog alerts announced in the media. For example, southern California's air quality agency announces stage 1, 2, and 3 alerts as ozone levels rise progressively above the federal safety limit. Stage 1 alerts are for people with heart or respiratory conditions, stage 2 for anyone exercising strenuously, and stage 3 for everyone (schools and businesses are closed). Stage 1 alerts are fairly common in cities with an ozone problem. Given children's special vulnerability to air pollution, parents may want to observe stage 1 alerts.

Next, consider restricting your children's vigorous exercise to times when or places where the ozone level is low. Generally, the ozone hazard is minimal in early spring, late fall, and winter. During the summer, ozone begins building during the morning rush hour from the combination of auto emissions, heat, and sun. The hotter the day, the more ozone. Wind carries it away from urban centers into suburbs, where the level can peak in the afternoon. Children should decrease their outdoor activity on hot, heavily polluted days. They can also do what joggers and cyclists do: exercise away from and upwind of roadways, especially highways. Try to keep them away from severely polluted areas between eleven o'clock and three. Plan family vacations in places that are less polluted.

Remember, too, that in well-ventilated homes without air-conditioning, indoor ozone levels are not negligible. Scientists at a Bell Communications Research facility in New Jersey recently found out that indoor ozone can reach levels higher than 70 percent of outdoor levels.

HOW TO KEEP YOUR FAMILY FROM POLLUTING

There are many practical changes each family can make in its routine that collectively help keep outdoor air clean. Through adopting new habits, parents are teaching their children by example.

Be a Green Consumer

Parents can learn and teach their children how to be "green" consumers, decreasing their use of products that contribute to air pollution. Here are some examples:

- Reduce your dependence on dry cleaners, which use and release toxic chemicals (see chapter 16, "Choosing Safer Household Products").
- Avoid gasoline-powered lawnmowers, which are not regulated for emissions.
- Try to buy paint and related products like varnish, glues, and deodorants that are formulated to contain fewer volatile solvents (see chapter 16).
- Avoid self-starter coals or lighter fluid in your grill. Try to light coal with kindling instead, or use a gelled alcohol product like Alco Brite™ (see Resources).
- Add bushes, hedges, and trees to your yard, which filter and keep clean air, and reduce greenhouse gases. Free trees and planting instructions are available from the National Arbor Day Foundation (see Resources).

Change Your Driving Habits

Like most Americans, you probably won't consider giving up your car to reduce air pollution. But you can change your driving habits to use less fuel and emit less exhaust. Children will learn that how and when the car is used is as important as safe driving. Here is a list of suggestions, some of which come from the Shell Oil Company (see Resources):

- Avoid solo driving, and instead use public transportation, carpools, and bikes.
- Do several errands in one trip.
- Avoid unnecessary idling. Starting the car again uses less gas than idling for 30 seconds or more.
- When you use the self-serve gas pump, don't spill gas, which evaporates and pollutes the air. Don't overfill or top off the gas tank; stop when you hear the nozzle click.
- Look for gas stations that have installed vapor-recovery nozzles on their pumps. You can identify them by the little holes at the end of the nozzle.
- Look for gas stations that sell reformulated gasoline, which doesn't evaporate as readily as ordinary gas and burns more cleanly.
- Avoid driving old cars that release a lot of dirty exhaust. Cars built before 1981 produce more than 80 percent of auto pollution.
- When you buy a new car, look for one with fuel efficiency and an improved catalytic converter.
- Use your air-conditioner as little as possible, as it uses more than a gallon of gasoline for each tankful you burn while driving.
- Keep your tires inflated so they will not cause drag, which can raise fuel consumption by 6 percent.
- Use radial tires, which improve fuel economy by about a mile per gallon.
- To burn less gasoline, drive at a steady speed, and slowly when traffic permits.
- Remember three maintenance steps: keep the car tuned, check the emissions system regularly, and replace your air filter at least every 15,000 miles.

RESOURCES

Products and Services

Baubiologie Hardware, 200 Palo Colorado Canyon Rd., Carmel, CA 93923. Telephone: (800) 441-8971 or (408) 625-4007.

Offers an alternative to petrochemical-based lighter fluid and cooking fuels called Alco Brite™.

National Technical Information Services, 5285 Port Royal Rd., Springfield, VA 22165. Telephone: (703) 487-4650.

Produces hard copy reports from the National Library of Medicine's TOXNET computer database.

Toxicology Information Program, National Library of Medicine, 8600 Rockville Pike, Bethesda, MD 20894. Telephone: (301) 496-6531.

Puts out information on how to get on-line with the TRI and TOXNET database.

Sources of Information

EPA. Emergency Planning and Community Right-to-Know hotline: (800) 535-0202. In Washington, D.C., call (202) 479-2449.

Request the EPA booklet "Chemicals in Your Community, A Guide to the Emergency Planning and Community Right-to-Know Act."

Island Press, Star Route 1, Box 38, Covelo, CA 95428.

Write away for *The New York Environment Book*, which discusses air pollution in New York City.

National Arbor Day Foundation, 100 Arbor Avenue, Nebraska City, NE 68410. Telephone: (402) 474-5655.

Promotes planting and care of trees. Offers free Colorado Blue Spruce trees, educational programs for grades K-6, the Conservation Trees program, and the Tree City USA project.

Natural Resources Defense Council, NRDC Publications, 40 W. 20th St., New York, NY 10011. Telephone: (212) 727-2700.

Write the NRDC to order *A Who's Who of American Toxic Air Polluters: A Guide to More than 1500 Factories in 46 States Emitting Cancer-Causing Chemicals*.

Shell Oil Company, P.O. Box 4681, Houston, TX 77210.

Write Larry Olejnik at this address for a free copy of the pamphlet "Protect Your Car and the Environment."

CHAPTER 8

Sunning Safely

The Problem. Man-made chemicals are thinning the ozone layer, allowing an increasing amount of dangerous ultraviolet radiation (UV) to reach the earth, where it can damage tissue like the skin and eyes.

The Risk to Kids. Children who are encouraged to play in the sun without protection and who get severely sunburned even a few times increase their risk of getting skin cancer, the most common kind of cancer. Children who don't wear protective sunglasses in very sunny conditions risk long-term damage to the eyes.

What to Do. Parents can learn about the risk factors for skin cancer, teach their children how to avoid sun overexposure, and choose for them an effective sunscreen and protective sunglasses. Parents can also buy products that do not contain ozone-depleting chemicals, and carefully service and dispose of air-conditioners and refrigerators that do.

Americans are getting more skin cancer than ever before. Experts cite two reasons: more sun exposure over the last three decades, and an increase in the amount of UV rays in sunlight. A large proportion of lifetime sun exposure comes before the age of eighteen, and it is now clear that people need sun protection during their early years. The story of a woman named Tracy, described by noted medical writer Dr. David Reuben in *The Reader's Digest*, illustrates how dangerous tanning habits have become. Tracy is a blond, blue-eyed, fair-skinned woman. She spent her childhood on sunny beaches and ski slopes. She visited a tanning parlor to look her best in her wedding gown. She honeymooned on the beach in Hawaii. About three years later, when she was twenty-five, a dark mole appeared on her back and she was diagnosed with malignant melanoma. She underwent surgery. Luckily, the tumor hadn't spread. But since then, twenty more suspicious moles have been removed from her chest, abdomen, and back. Tracy's experience means she will protect her children from excessive sun exposure. But all her efforts cannot stop the harmful UV radiation in sunlight. That is the job of "good" ozone, the ozone of the stratosphere.

GOOD OZONE

Good ozone is distinguished from smoggy and irritating ground-level ozone by its location. Good ozone sits in the stratosphere, the layer of atmosphere 10 to 30 miles above the earth. There it shields the earth's surface from most of the dangerous UV radiation in sunlight. Life on earth evolved over millions of years under conditions that included a fairly constant UV level. It now appears that man-made pollution has—in less than a century—begun depleting the ozone layer. As a result, more UV radiation is reaching the earth. That life forms are sensitive to the change is clear from the death of certain species of sea phytoplankton, the basis of the ocean's food web, and from damage to land crops and increased rates of skin cancer in humans.

WHY IS THE OZONE LAYER SHRINKING?

Stratospheric ozone is shrinking because chlorine from man-made chemicals is breaking apart its molecules. These chemicals, which include methyl chloroform, carbon tetrachloride, and chlorofluorocarbons (CFCs), are industrial solvents. Some are used in consumer products. For example, CFCs, nonflammable, noncorrosive, and nontoxic, have been used in refrigerator coolant, aerosol propellants, cleaning solutions for electronic parts, and styrofoam.

The stability of these solvents is what causes the problem. Once they evaporate, they persist, some for as long as 100 years. That is more than enough time for them to drift 20 miles up to the stratosphere, where they are finally broken up by UV bombardment. The liberated chlorine in turn breaks up ozone.

By 1973, millions of tons of CFCs had been produced, but it wasn't until 1984 that the first huge ozone hole in the stratosphere was recorded over the Antarctic. These dramatic holes appear around September when extremely cold air gathers in the polar stratosphere and speeds up the chlorine destruction of ozone. The same destruction process goes on around the globe but at a slower rate. The EPA announced in 1991 that ozone losses across the middle of the United States (Seattle to New Orleans) are worse than previously thought, up to 5 percent per decade. Moreover, the relatively high losses that occur during the winter months are dragging into April and May, when children return outdoors to play for the summer.

It is disconcerting that efforts to stop production of ozone-depleting chemicals won't immediately stop the thinning of the ozone layer. One problem is that CFCs now in the atmosphere will last for many decades. The other is that during the next few decades, some 20 billion pounds of ozone-depleting chemicals will be released into the atmosphere. Worldwide CFC production won't end until the year 2000. The biggest reservoir in the world of still to be released CFCs sits in the refrigerators and air-conditioners of America.

WHY CHILDREN ARE VULNERABLE TO UV RADIATION

Children are encouraged to spend lots of time in the sun as part of a healthy lifestyle. They begin absorbing UV radiation as soon as they are introduced to the sun. The effects of this radiation, a major cause of skin cancer, are cumulative. There is a lag time of twenty to thirty years between exposure and apparent damage. The statistics on skin cancer in young adults show how the danger is growing. The number of cases of squamous cell carcinoma increased 260 percent in men and 310 percent in women from 1963 to 1990. The incidence of malignant melanoma, the deadliest of skin cancers, increased even more in the same period. Dermatologists are calling melanoma a near-epidemic cancer, and some predict that by the year 2000, one in ninety Americans will get it.

The three most common types of skin cancer are basal cell carcinoma, squamous cell carcinoma, and malignant melanoma. Basal cell carcinoma is the most common, especially on the neck and face. It usually grows slowly and stays localized. Squamous cell carcinoma tends to be more aggressive, growing faster and spreading more. Malignant melanoma is the most dangerous. It may spring from an existing mole, but not always. Malignant melanoma must be diagnosed and surgically removed or else it will spread from the skin to other parts of the body.

Recent research by Dr. Darrell Rigel at New York University reveals that people at highest risk for developing melanoma have six roughly equal risk factors: blond or red hair, marked freckling of the upper back, family history of melanoma, a precancerous skin condition called actinic keratosis, three or more blistering sunburns before age twenty, and an outdoor job for three or more summers as a teenager (without sun protection). High-risk youth (those with light or red hair, for example) should avoid too much sun and know how to protect themselves. People with black skin, whose pigment provides a natural kind of sun protection, have a much lower incidence of skin cancer.

There are other consequences of exposure to UV radiation: premature aging of the skin (UV rays destroy the elastic tissue fibers that keep skin taut and supple), eye damage (UV rays turn the eye's clear lens a brownish color, making a person susceptible to cataracts and partial blindness), and immune system damage (UV rays kill immune cells that fight off disease).

WHAT IS THE GOVERNMENT DOING?

In 1978, the EPA banned nearly all CFCs in aerosol products. Accumulating evidence of ozone depletion spurred an international effort to control the CFC hazard. In 1987, under the provisions of the Montreal Protocol, major industrialized nations pledged to cut CFC production in half by 1999 and to freeze production of halons, a group of even more powerful ozone-depleting chemicals used mainly in fire extinguishers. Still more evidence of ozone depletion led to the 1990 London Treaty amending the Montreal Protocol. In this pact, signatories agreed to phase out the production and use of all CFCs by the year 2000, and of methyl chloroform (another ozone depleter) by 2005. (Many Western nations—Australia, Canada, and eleven others, but *not* the United States—will eliminate CFCs even before 2000.) Other significant features of the treaty are new controls on the use of carbon tetrachloride and halons, a $240 million fund to help developing nations eliminate CFC use, a plan to phase out hydrochlorofluorocarbons (HCFCs), which are initial replacements for CFCs but not completely safe, between 2020 and 2040, and an agreement to renegotiate the treaty in 1992.

Enforcement of the London Treaty may prove difficult. Chemical and appliance manufacturers in dozens of countries will need to be monitored for compliance. The situation is critical in the United States, which is the biggest producer and user of CFCs in the world. The reservoir of not-yet-emitted CFCs and halons in existing refrigerators, air-conditioners, and fire extinguishers should be captured and recycled. The Natural Resources Defense Council (NRDC), an environmental group working since the

1970s to protect ozone, suggests that the EPA mandate CFC recycling in domestic appliances and commercial equipment. Government-approved standards are important to small businesses like auto service stations. They don't want to invest thousands in recycling equipment only to find out later it doesn't meet government specifications.

Meanwhile, the use of HCFCs as replacements for CFCs has prompted criticism from environmentalists. HCFCs are about one-fiftieth as ozone-destructive as CFCs, but massive increases in HCFCs could erase this advantage.

HOW CHILDREN CAN SUN SAFELY

A reasonable amount of sun is good for people of all ages. Bodies need sun to produce vitamin D, which is present in only a few foods, for example, fortified milk, butter, and egg yolk. But children need only a little sun to satisfy their vitamin D requirement—10 to 15 minutes between eleven and two on summer days, two to three times a week. Children should learn sensible sunning habits to avoid overexposure. And because the ozone layer is likely to continue thinning for decades, they need protection from increasing UV radiation.

Avoiding Overexposure

The basic rules for sun exposure can be borrowed from Australia, where the increase in skin cancer has been the most dramatic. Why does Australia have the highest skin cancer rate in the world? Part of the answer is that many fair-skinned northern people have migrated to this sunny country. People with red or blond hair and blue eyes have less protective pigment. Moreover, UV radiation over Australia has increased dramatically. In any case, the Australian approach is "slip, slop, and slap." This means:

- Slip on protective clothing. The tighter the weave (preferably cotton), the greater the protection.
- Slop on sunscreen to protect exposed skin. Use sunscreen even on cloudy days—up to 75 percent of UV rays pass through clouds. (Clouds stop more of the warm-feeling infrared rays.)
- Slap on a hat to protect the head and face. A baseball cap keeps about half of dangerous UV rays out of the eyes. Also protect the eyes with UV-filtering sunglasses.

In America, the National Institutes of Health (NIH) adds the following recommendations to the advice above:

- Use a sunscreen that is broad spectrum and waterproof, and has a sun protection factor (SPF) of 15.
- Minimize exposure to the sun between ten and three, when the sun's rays are most intense.
- Consult with a physician about medications that can increase sensitivity to UV light.

Finally, the FDA cautions parents not to apply sunscreen to babies until they are six months old.

How to Choose Sunscreen

Some sun protection products like zinc oxide ointment are physical barriers that reflect or scatter all light. Other sunscreens are chemical barriers that absorb ultraviolet light. Sunscreen for your child should be strong enough to offer adequate protection yet not irritate young skin.

Broad-spectrum sunscreens are a must, and most products display this property clearly on the label. They filter out two kinds of dangerous UV radiation called UVA and UVB. UVB rays have shorter wavelengths and are the principal cause of sunburn, skin cancer, and premature aging of the skin. UVA rays have longer wavelengths and penetrate more deeply into the skin. They contribute to skin cancer and aging. A nonwaterproof sunscreen should be applied about one-half hour before your child goes in the sun, so that the sunscreen can saturate the skin layers needing protection. The recommended amount of sunscreen for adults is one fluid ounce; reduce the amount depending on the size of your child. Don't forget to put sunscreen on all exposed skin. Frequently missed areas are the lower back, thighs, and ears.

Sun protection factor (SPF) numbers indicate how long a person can stay in the sun before he or she begins to burn. If you use an SPF 15 sunscreen, you can stay outdoors fifteen times longer before burning than if you had used no sunscreen. The average fair-skinned person begins to burn after twenty minutes in full sun. Wearing the SPF 15 sunscreen allows that person five hours in the sun before burning. Remember that reapplication of sunscreen doesn't extend the time of protection, but helps maintain the protection. The SPF system measures only UVB, not UVA, radiation protection. Most sunscreens contain several active ingredients, one or more of which provides some UVA protection.

All sunscreens use the same FDA-approved list of active ingredients. Common ones are cinnamates, salicylates, and para-aminobenzoic acid (PABA), which provide UVB protection; Parsol (avobenzone), which provides UVA protection; and benzophenones, which protect—to a limited extent—against both UVA and UVB. PABA has been around for fifty years and is considered safe. Most children's sunscreens omit PABA because it can cause skin irritation. One former sunscreen ingredient, urocanic acid, has been dropped from American—but not European—products because it has been linked to skin cancer in experimental mice studies. Of the ingredients that provide UVA radiation protection, only one, Parsol, has been officially approved by the FDA as a safe and substantially effective UVA screen. At present it is available only in Filteray (from Burroughs Wellcome) and Photoplex (from Herbert Laboratories) sunscreens. Some commercial sunscreens are advertised as "natural." Most of these products have the same group of FDA-approved active ingredients; they just offer natural inactive ingredients like vegetable oil.

Also on the market are "bronzers," or self-tanning lotions. These contain dihydroxyacetone (DHA), which the FDA considers safe as a skin dye. Bronzers don't offer sunscreen protection. Dermatologists consider tanning parlors and sun lamps unsafe for skin, regardless of what kind of UV rays they generate. Parents and older siblings can set a good example for the very young by avoiding tanning parlors and sunlamps completely.

How to Choose Protective Sunglasses

Light dangerous to the eyes includes UVA and UVB rays as well as blue light, from the blue end of the visible light spectrum. Animal studies suggest that exposure to UVA and blue light contributes to macular degeneration, a disease of the retina causing loss of central vision. At present, there is no federal regulation of UV- or blue light–protection levels in sunglasses, and no law requiring consumer information labels. Lens darkness, color, and price don't indicate the quality of UV protection. Some manufacturers have begun to use the voluntary labeling developed by the American National Standards Institute (ANSI), a private organization that sets standards for a wide range of consumer goods. ANSI divides sunglasses into three categories: cosmetic, general purpose, and special purpose. Cosmetic sunglasses block the least UV radiation, special purpose block the most. The label states both the category and the minimum blockage required in that category. The ANSI standard is strictest for UVB protection. For example, general-purpose sunglasses block 95 percent of UVB, but only 60 percent of UVA. If a general-purpose model carries a little more protection than the minimum necessary, this is also stated on the label. Some authorities feel the voluntary labeling program is inadequate.

Parents should check for ANSI labels when buying sunglasses, choosing those that have good UVA and UVB protection. Sunglasses providing UVA and UVB protection absorb wavelengths from 280 to 380; those providing blue light protection absorb wavelengths from 280 to 500. Amber-colored (yellow) and brown glasses block out blue light better than other lenses. Another option parents have is to get their children's sunglasses chemically dipped by an optician. The cost will probably be about $25. Your optician may have equipment to test any sunglasses your child already has for their degree of UV protection. If your children play sports, try to find them polycarbonate protective goggles with an anti-scratch coating.

Children's sunglasses should block at least 75 percent of visible light. You should not be able to see your eyes in a mirror when wearing the sunglasses. If your child is out in extremely bright conditions like the ski slope or the beach—where light is reflected from snow, sand, and water—choose the darkest, "special purpose," models.

WHAT YOU CAN DO TO PROTECT THE OZONE LAYER

The environmental organization Greenpeace urges citizens to take action against companies in their communities producing and using ozone-depleting chemicals. Citizens can consult the EPA's TRI (see Resources) to find out if a local company is releasing ozone-depleting chemicals. The NRDC publishes the *Who's Who of American Ozone Depleters* (see Resources). Citizens can ask their government representatives to oppose chemical releases and to support legislation curbing them. Another way to help protect ozone is through consumer habits.

Change Your Purchasing Habits

First, try not to buy aerosol products that escaped the 1978 ban on nonessential CFCs. Examples are spray confetti, VCR head cleaners, negative film cleaners, boat-warning horns, car air-conditioner refills, and drain plungers. Look for alternative products with labels that state they contain no CFCs. Avoid styrofoam cups and plates (some of these are made with the safer HCFC chemicals, but they are not entirely ozone-friendly or biodegradable). Observe the local ban on polystyrene foam, should it exist.

How to Service and Dispose of CFC Equipment

People can try to have coolant recycled when the air-conditioner in their car is serviced. Ninety million vehicles carry coolant, the disposal of which accounts for an estimated 16 percent of ozone destruction. According to the Mobile Air-Conditioning Society, recycling equipment—called vampire

machines because they suck out coolant—became available in shops in
1990. Avoid do-it-yourself cans to replenish coolant. Refilling a leaky
system only adds more CFCs to the atmosphere.

The EPA hopes to mandate recycling programs for refrigerator coolant
soon. Until then, keep your refrigerator working. Ask your appliance
dealer when and if models with CFC-free coolant and insulation will be
available. Also, hold onto your halon fire extinguisher until a halon
reclaiming program is established. New buyers have the alternative of
purchasing a dry-chemical extinguisher, which is adequate for homes (see
Resources).

RESOURCES

Products and Services

Baubiologie Hardware, 200 Palo Colorado Canyon Rd., Carmel, CA 93923.
Telephone: (800) 441-8971 or (408) 625-4007.

> Offers a free catalogue on multipurpose dry-chemical fire extinguishers that
> won't harm the ozone layer.

Toxicology Information Program, National Library of Medicine, 8600 Rockville
Pike, Bethesda, MD 20894. Telephone: (301) 496-6531.

> Gives out information on computer access to the TRI. Also check the library
> for microfiche copies of the TRI.

Sources of Information

American Academy of Dermatology, P.O. Box 3116, Evanston, IL 60204-3116.
Telephone: (708) 869-3954.

> Send a self-addressed, stamped envelope to this address for the free brochure
> "Sun Protection for Children."

American Academy of Pediatrics, Dept. C, P.O. Box 927, Elk Grove Village, IL
60009-0927. Telephone: (708) 228-5005.

> Send a self-addressed, stamped business-sized envelope for a free copy of the
> brochure "For Every Child Under the Sun: A Guide to Sensible Sun Protec-
> tion."

American Cancer Society. Telephone: (800) ACS-2345.

> Call this number for the ACS office nearest you. The ACS offers a free
> brochure, "The First Number to Teach Your Children Is Number 15," and a
> children's activity wall poster called "Sun Protection."

American Optometric Association, Communications Center, 243 N. Lindbergh
Blvd., St. Louis, MO 63141. Telephone: (314) 991-4100.

> Send a self-addressed, stamped envelope for two free pamphlets, "Are Your
> Eyes Safe from UV Radiation?" and "Sunglasses Are More than Shades."

Natural Resources Defense Council, Publications Department, 40 West 20th St., New York, NY 10011.

Publishes and distributes *A Who's Who of American Ozone Depleters* and *Saving the Ozone Layer.*

"Sunscreens." *Consumer Reports* 56 (1991): 400-406.

Provides a rating of sunscreen products.

Part 3

The Healthy Meal

Good nutrition is fundamental to a child's health. Parents begin making choices about nutrition from the very beginning, when they decide whether their baby will be breast- or bottle-fed. From that time on, the choice of food depends on what's healthy, what's economic, what's convenient, and what the child likes. It would help if parents could rely on what nutritionists consider to be healthy foods. But today there are concerns about the residue of cancer-causing pesticides in fruits and vegetables, bacterial contamination of chickens and eggs, and synthetic hormones and drugs in milk. There are also questions about the safety of food additives and drinking water contaminants. This section evaluates food and water hazards and examines ways to reduce your children's exposure to them.

CHAPTER 9

Purging Pesticides from Produce

The Problem. The average preschool diet of fruits and vegetables exposes a child to levels of neurotoxic and carcinogenic pesticides that the EPA considers unsafe.

The Risk to Kids. Children are especially vulnerable to carcinogens and neurotoxins. The latter can act on the nervous system at critical stages of development, resulting in irreparable motor, behavioral, and cognitive damage.

What to Do. Parents can minimize their children's exposure to pesticides by buying organic, seasonal, and domestically grown produce, thoroughly washing or peeling it, or growing their own organically.

In 1989, the TV show "60 Minutes" aired a story about a highly respected environmental group that pointed the finger at the carcinogen alar in apples. With each bite of the fruit, it warned, children were being poisoned. Apples were removed from some school cafeterias, consumers boycotted apples and applesauce at the supermarket, parents were seized with new doubts about the safety of produce. Today, the danger of pesticides is still a concern for many parents as well as a source of controversy among legislators and scientists.

WHY ARE PESTICIDES USED?

A third of the world's food crop is lost every year to thousands and thousands of diseases, weeds, worms, and insects. To protect their crops from this formidable onslaught, American farmers spend around $15 billion a year on

800 million pounds of pesticides. To some extent, these chemicals have made possible our bountiful harvest. The dark side is that they may explode into cancer or neurological problems sometime during our children's lifetime. Pesticides contaminate not only fruits and vegetables but also meat from animals that graze on treated crops and fish from waters polluted by agricultural runoff.

WHAT IS THE GOVERNMENT DOING?

Government regulation of pesticides has three basic functions: to set requirements for the registration of pesticides, including criteria for protecting human health and the environment; to establish limits or "tolerances" for the amount of pesticide residue that can remain on crops going to market, and to monitor residues in crops, ensuring that such limits are not exceeded.

One of the biggest problems with government regulation is the lack of good data on the health effects of pesticides. Many older pesticides are suspected of being harmful to humans based on the limited knowledge that does exist. The EPA itself has stated that there is evidence of carcinogenicity for 66 or more of the approximately 360 pesticides used on foods. A 1987 National Academy of Sciences study estimated that 90 percent of all fungicides, 60 percent of all herbicides, and 30 percent of all insecticides may be carcinogenic.

Congress recognized the urgent need for reevaluation of older pesticides. In 1972, it amended the Federal Insecticide, Fungicide, and Rodenticide Act (FIFRA), requiring reevaluation and reregistration of all pesticides in light of up-to-date knowledge and testing methods. For the first time, older pesticides were to be scrutinized for their chronic health effects as well as environmental impact. Unfortunately, only a handful of pesticides have been reregistered, and there are still major data gaps on the effects of many pesticides. The FIFRA amendments passed by Congress in 1988 should expedite the process of data gathering and reregistering; the EPA is required to complete the job by 1997.

Even if the EPA could complete its retesting of all pesticides by 1997, the real impact on public health may well be determined by what risk-taking criteria the EPA uses for reregistration. Three different criteria have been used: the "risk-benefit" standard, the "zero-risk" standard, and the "negligible-risk" standard.

The risk-benefit standard, introduced in the 1972 FIFRA amendments, permits registration and tolerance-setting of a pesticide if it poses "no unreasonable adverse effects to man or the environment," while simultaneously "taking into account the [pesticide's] economic, social and environmental costs and benefits." The zero-risk standard, enacted as the Delaney Clause of the Food, Drug and Cosmetic Act in the 1950s, prohibits any residues of cancer-causing chemicals in food after processing. The negligible-risk standard, a new interpretation of the Delaney Clause, permits registration and tolerance-setting for a pesticide only if the health risks from cancer are lower than a level considered to be insignificant. The EPA's 1988 adoption of this standard is now being legally challenged by environmental groups.

These three standards differ greatly in the degree of risk considered acceptable. Naturally, environmental groups and the pesticide industry are at odds over them. Environmental groups favor the zero-risk standard, whereas industry prefers the risk-benefit approach.

Major inconsistencies in federal policy have resulted from the way the EPA applied these standards in the past. For example, some sixty-six older pesticides that the EPA itself admits are carcinogenic in test animals are still legally allowed on food crops. This is because the EPA never enforced the zero-risk standard, choosing instead to apply the risk-benefit criterion. These older pesticides are believed to contribute up to 90 percent of the total cancer risk from all pesticides in the diet. However, since 1978 the EPA has applied the zero-risk standard in its review of newer, often less hazardous pesticides. This has lead to the so-called Delaney Paradox in which some older, very toxic pesticides are allowed to remain on the market while new, less toxic ones may not be granted registration.

In an attempt to rectify such inconsistencies, the EPA recently adopted a negligible-risk standard for old and new pesticides alike. According to the policy, a pesticide will be approved if its residues in processed food do not cause more than one additional cancer per million persons exposed over a lifetime. However, the pesticide could still be registered for use in raw produce if the benefits outweigh the risks (risk-benefit standard). Unfortunately, by allowing the risk-benefit standard for raw fruits and vegetables, these seemingly healthy foods may be tainted with pesticides.

THE RISK TO CHILDREN

Probably the best data on the risk to children is the highly acclaimed 1989 study by the Natural Resources Defense Council (NRDC), "Intolerable Risk: Pesticides in Our Children's Food." It concluded that between 5,500 and 6,200 of the current 22 million preschoolers will get cancer sometime in their life from exposure to just eight pesticides that are legally permitted in foods. The major risk by far was from alar, a chemical that prevents premature ripening of apples. Alar penetrates apples and can't be washed off. The NRDC calculated a risk of 1 cancer case for every 4,200 preschoolers exposed to alar during the first six years of life. This is a whopping 240 times greater than the risk considered acceptable by the EPA. Because of pressure from environmental groups and the public, production of alar for food crops was voluntarily canceled by the manufacturer.

The NRDC study found that amounts of four of the commonly used carcinogenic fungicides in fruits and vegetables would result in a risk of 1 cancer case for every 33,000 to 160,000 children exposed. In addition, it found that at least 17 percent of the preschooler population is continually exposed to levels of neurotoxic pesticides that exceed EPA standards. These levels can cause a wide range of health problems: motor (loss of coordination, paralysis), behavioral (irritability, depression), and cognitive (speech, learning, and memory impairment).

The NRDC believes that preschoolers' risk from pesticides is probably even greater than its study predicted. The study examined the risk to children through the age of five, not the risk from the age of six and up. Only 23 pesticides out of the 360 registered by the EPA for food crops were included, and only 8 of the 66 pesticides believed to be carcinogenic were evaluated.

The NRDC looked at fruits and vegetables, not the other food items consumed by preschoolers that can be contaminated with pesticides. Take milk, for example. The EPA has estimated that most of the cancer risk from the fungicide captan comes from milk. Drinking water may also be a substantial source of pesticides. The EPA has reported quantities of at least seventy-seven pesticides in the groundwater of thirty-nine states. Nor did the NRDC study consider the possible effects of multiple pesticide exposure in the diet. Very little data is available in this area.

The NRDC did not consider the toxicity of inert ingredients (solvents and stabilizers like formaldehyde and alcohol) that can make up as much as 99 percent of a pesticide compound. Of 1,200 substances classified as inert, the EPA judges that 55 cause adverse health effects such as cancer or birth defects in animals and humans. No health information exists for some 800 inert ingredients; another 200 appear to be harmless. Manufacturers are not legally required to list inert ingredients, considered "trade secrets," on pesticide packaging. In 1987, the EPA began encouraging the removal of the most hazardous inerts from pesticides, but no formal policy has been adopted.

Another factor that may have limited the NRDC risk estimates is that data on pesticide residues in food was taken from the FDA, which is responsible for enforcing tolerance levels. FDA tests detect only about 40 percent of the pesticides that may leave residues, and some of the undetectable pesticides are the most hazardous. Unfortunately, the FDA lacks the resources to do a better job. In a given year, it tests about 1 percent of imported produce, less than 1 percent of domestically grown produce, and even less of processed products like apple juice.

Why Are Children More Vulnerable?

Children face a greater risk from pesticides than adults for both behavioral and physiological reasons. Children's exposure is greater because they consume many more vegetables and fruits than the average adult both in absolute numbers and also relative to their body weight. The average child eats seven times more apples and drinks eighteen times more apple juice than the average adult. However, the EPA sets tolerances based on what adults consume, ignoring the consumption patterns of children. This means the concentration of pesticides on foods consumed in great quantities by children, though legal, may be unsafe. The NRDC estimates that half of the lifetime risk of cancer from pesticides in the diet is incurred by the age of six.

There are also physiological reasons children are more vulnerable to pesticides (see chapter 5, "Maintaining a Chemical-Free Yard"), including the fact that their developing brain and nervous system can be irreparably damaged by neurotoxic pesticides.

PESTICIDE RISKS IN PERSPECTIVE

Industry representatives and some scientists feel the risks posed by synthetic (man-made) chemicals in the diet are being exaggerated, and that the focus on these chemicals diverts energy and funds from more important health threats like smoking and fat consumption.

The Natural Pesticide Controversy

Bruce Ames, a well-known research scientist at the University of California, has sparked an enormous debate among scientists and regulators regarding the relative risks of synthetic pesticides and natural pesticides. All plants, including food crops, contain toxic chemicals that provide protection against predators like insects, fungi, and animals. Food crops are also frequently contaminated with molds that produce toxic chemicals.

According to Ames, we ingest 10,000 times more natural pesticides by weight than man-made ones.

Some of the natural pesticides in foods like peanut butter, brown mustard, basil, and mushrooms have been shown to be carcinogenic in animals or humans. Some, like aflatoxin, a contaminant of corn, peanuts, and wheat, have long been known for their toxicity. Aflatoxin is an antibiotic produced by molds to kill off competing species. It is associated with human liver cancer, especially in tropical countries where aflatoxin is common. According to Ames, one peanut butter sandwich is a hundred times more hazardous to the health than the average daily intake of several synthetic pesticides he has studied.

Although Ames's theory has been welcomed by industry and regulators, there are criticisms. Only a few dozen natural plant chemicals have been tested for toxicity. Of these, approximately half showed some evidence of carcinogenicity. However, these chemicals may have been chosen for study in the first place because they were suspected of being carcinogens. At this point, there simply isn't enough data on natural carcinogens to determine how much of a risk they are to human health.

Even if natural pesticides in foods are a greater health hazard than synthetic pesticides, should synthetic hazards be dismissed as trivial and efforts to regulate them abandoned? One of the benefits of this controversy is to spotlight the process by which our society regulates pesticides. This should improve the regulation of all toxic chemicals.

The Animal-Testing Problem

Another criticism that Ames and some other scientists have with government regulation of pesticides is that it is based on the results of testing in which high doses of pesticides are fed to animals. About half of all chemicals tested in animals have turned out to be carcinogens. Ames believes that in many of the tests the resulting cancer is an artifact of exposure to high doses. High doses are used for economic reasons; less time and fewer animals are needed to see a health effect. Studies have shown that high doses of many chemicals can cause cell death. To repair the damage, remaining live cells proliferate, increasing the chances of a carcinogenic mutation in their DNA. Thus, many chemicals may be carcinogens only when administered at high doses. Ames argues that results of animal testing at high doses cannot predict the health effects of chemicals in humans where exposure levels are hundreds, or even thousands, of times lower.

Other scientists don't subscribe to Ames's arguments. They are concerned that the focus on cell proliferation oversimplifies the carcinogenic process, which is known to involve multiple factors. Critics also cite the fact that extensive cell proliferation during normal bodily processes like skin replacement does not cause cancer.

The controversies over natural carcinogens in food and high-dose animal testing point out the critical need for greater understanding of how chemicals cause cancer and for improving the process by which the safety of pesticides is determined.

WHAT CAN A PARENT DO?

Given the conclusions of the NRDC study and the problems with government regulation and animal testing, it seems prudent to minimize your children's exposure to pesticides while continuing to provide them with a varied diet. There are several ways to do this:

• Buy organically grown fruits and vegetables. Many large supermarket chains now have an organic food section. Prices are often higher and supplies fewer. Make sure you are getting the real thing by discussing the source of the food with your grocer. You might also look for small farms in your area that grow organic produce, or for food-buying clubs that have arrangements with organic growers. Mail order is another source for organic produce as well as meats, cereals, and dairy products. Contact the Americans for Safe Food or the California Action Network for lists of suppliers (see Resources).

What food qualifies to be called organic? The Organic Foods Production Act of 1990, which is part of the 1990 Farm Bill, stipulates that only those foods grown on farms that have been herbicide-, fungicide-, and insecticide-free for at least three years can be marketed as organic. State agricultural departments and private certification agencies will be charged with enforcing the law. Until such foods appear in the market, perhaps in 1993, the consumer must be cautious of food purported to be organic. Your best protection is to inquire about the foods you buy and to purchase only foods marked "certified organic." Currently, independent certifiers, or the states themselves, specify the requirements that must be met before crops are labeled certified organic. In some states, foods are certified organic if they have been grown in soil without chemicals for one year; in other states, the requirement is three years. If you wish to know how the food was grown, write to the certifying organization if one is listed on the package. For a complete list of certifiers throughout the country, send a self-addressed, stamped envelope to the Organic Foods Production Association of North America (see Resources).

• Grow your own produce. See chapter 5, "Maintaining a Chemical-Free Yard," for suggestions on growing crops without pesticides.

If you must buy foods that are not organically grown, the following tips should help:

• Buy produce in season. Out-of-season produce is often imported and may contain either a high level of pesticides or pesticides that have been banned from use in the United States. Your grocer should be able to provide the country of origin of any produce.

• Look for supermarket produce that has been privately tested for pesticides. A growing number of supermarkets are now contracting with private pesticide-testing labs to certify that their regular produce contains negligible residues. Private companies like Nutriclean of Oakland, California, have filled a void created by the government's inability to effectively monitor pesticides in our food supply. The stamp of approval from these companies means that the food has met the EPA's negligible-risk standard.

Critics of this growing industry argue that only about 1 percent of the produce tested by private labs is shown to violate government standards, raising the question of whether such testing is only creating unwarranted fears among consumers and undermining the government's role in protecting the food supply. It is too early to tell whether private testing is just a fad or here for the long term.

• Beware of produce likely to have high levels of pesticides. Produce like strawberries and peaches have high levels of pesticides to preserve their appearance. In general, the more perfect-looking the produce, the more likely pesticides were used. Also, foods with edible portions grown in the soil—carrots, potatoes, celery—may readily absorb pesticides. Pesticide levels are often higher on leafy vegetables like spinach and lettuce because of their large surface area. A good source of information on specific pesticides that contaminate produce and whether washing reduces them is *Pesticide Alert*, available through the NRDC (see Resources).

• Peel waxed fruit and vegetables. Waxes are added to some twenty food items, including cucumbers, apples, bell peppers, tomatoes, melons, eggplants, grapefruit, oranges, lemons, limes, peaches, pumpkins, and sweet potatoes. Waxes, which prevent moisture loss and retard shriveling, are probably safe, but they have never been rigorously examined for health effects. The problem is they are often mixed with hazardous fungicides to prevent rotting. And, since they can't be washed off, they seal in other pesticides. Federal law requires that stores have signs on waxed produce. However, you seldom see them; the FDA is too overloaded to systematically enforce the law. Try persuading your supermarket manager by pointing out that failure to comply with the law carries a $1,000 penalty or a year in jail.

• Thoroughly wash produce or peel it. The best way to wash is with a brush in warm water. If you want to use a drop or two of mild dishwashing detergent, be sure to rinse thoroughly. It may be best not to wash leafy vegetables with soapy water, as they are difficult to rinse.

Peeling is the most effective way of removing nonsystemic pesticides, those that concentrate mainly on the peel. A disadvantage is that you are peeling away a good source of fiber. It is a good idea to discard the outer leaves on lettuce, cabbage, and other leafy vegetables, for these are the parts most likely to be contaminated with pesticides.

• Avoid the zest of the lemon, orange, lime, or grapefruit. The zest is the outer, colored part of the peel that contains intensely flavored citrus oils. Cooks often add finely grated portions of the zest or rind to enhance the flavoring of certain recipes. Certain oil-soluble pesticides may concentrate in the rind. Look for organically grown citrus fruit if you must use zest.

• Cook foods to reduce pesticide levels. Cooking destroys heat-sensitive pesticides that contaminate foods like meat, poultry, and fish. (It also destroys harmful microbes; see chapter 10, "Reducing Hazards in Poultry, Fish and Milk"). Steaming is a good way to cook vegetables if you want to minimize the destruction of vital nutrients.

• Trim fat from meat, fish, and poultry, and discard fat from broth and pan drippings. This will do away with pesticides that concentrate in fat.

• Don't pick wild berries or other plants growing next to roads, farm fields, or utility lines. These areas may have been sprayed with toxic herbicides. Instead, hunt for wild plants in deep woods or uncultivated fields.

• If you hunt or fish, consult with fish and game officials to determine if pesticides have been heavily used in the area. Many wild species eat contaminated food and build up pesticides in their tissues.

• Feed your infant commercial baby food rather than homemade food if the latter is made from nonorganic fruits and vegetables. Baby food manufacturers have reduced pesticides in their products. For example, the H. J. Heinz Company is avoiding fruits and vegetables containing residues of pesticides under review by the EPA.

RESOURCES

Sources of Information

Americans for Safe Food, Center for Science in the Public Interest, 1875 Connecticut Ave., NW, Suite 300, Washington, D.C. 20009. Telephone (202) 332-9110.

Will send you a list of mail-order organic food suppliers.

The Bio-Integral Resource Center, P.O. Box 7414, Berkeley, CA 94707. Telephone: (415) 524-2567.

Provides information on growing foods using IPM, and has a good publications list and products catalogue.

California Action Network, P.O. Box 464, Davis, CA 95617. Telephone: (916) 756-8518.

Write away for the *1991 Organic Wholesalers Directory and Yearbook*, a list of organic food distributors ($29.95 plus shipping).

The National Coalition against the Misuse of Pesticides, 701 E St., SE, Suite 200, Washington, D.C. 20003. Telephone: (202) 543-5450.

Provides information on pesticides and safer alternatives.

National Pesticide Telecommunications Network. Telephone: (800) 858-7378.

This hotline, funded by the EPA and Texas Tech University, answers questions about the toxicity of pesticides.

Natural Resources Defense Council, 40 West 20th St., New York, NY 10011. Telephone: (212) 727-2700.

Offers the following publications, among others: "Intolerable Risk: Pesticides in Our Children's Food," (summary, 1989), "Harvest of Unknowns: Pesticide Contamination in Imported Foods," "For Our Kids' Sake: How to Protect Your Child against Pesticides in Food," and *Pesticide Alert*.

Organic Foods Production Association of North America, P.O. Box 1, Belchertown, MA 01007. Telephone: (413) 774-7511.

Supplies lists of organic food certifiers and retailers.

The U.S. Public Interest Research Group, 215 Pennsylvania Ave. SE, Washington, D.C. 20003. Telephone: (202) 546-9707.

Will send you a copy of "Presumed Innocent: A Report on 69 Cancer-Causing Pesticides Allowed in Our Food," (10).

CHAPTER 10

Reducing Hazards in Poultry, Fish, and Milk

The Problem. Poultry, fish, and milk, considered some of the healthiest foods we feed our children, can be contaminated with harmful bacteria, industrial pollutants, and traces of drugs.

The Risk to Kids. The immature immune system and intestinal flora of young children leave them more vulnerable to bacterial infection from contaminated food. And their developing organs are more vulnerable to carcinogens that contaminate fish and milk.

What to Do. Parents can continue to feed their children these foods but should reduce the health risk by proper cooking and refrigeration and by omitting or reducing the amount of certain fish in the diet.

The diet of the American family has been steadily improving. The message that too much saturated fat and cholesterol are bad for the heart has resulted in the increasing popularity of poultry products and fish over red meats like steak. However, as good as they are for reducing fat and calorie intake, poultry and fish can be highly contaminated with bacteria and toxic industrial chemicals. This is partly a result of the frantic production needed to keep up with consumers' insatiable demand.

MANAGING MENACING MICROBES IN MEAT

Bacteria are clearly public enemy number one when it comes to food-related illnesses. According to the Center for Disease Control, during the five-year period from 1983 to 1987, bacteria caused 66 percent of foodborne disease outbreaks. Chicken has become the leading carrier of harmful bacteria. This is because chicken is now the most popular meat in the American diet. In fact, 1990 was the first year that Americans ate more chicken than beef or pork.

The Microbes

Illness from consumption of poultry is largely due to two different types of bacteria, salmonella and campylobacter. They contaminate anywhere from 40 to 70 percent of the 40 million chickens processed every year in the United States. Salmonellosis, the disease caused by ingesting chicken tainted with salmonella, accounted for 57 percent of bacterial foodborne disease outbreaks reported between 1983 and 1987. And reported cases represent only a fraction of the annual number of infected individuals, estimated to be in the millions. Campylobacter, little known or appreciated until recent years, is believed to cause illness in a couple million people every year.

There are two major factors for the rise in illness caused by salmonella and campylobacter. The first is the burgeoning demand for poultry. We just can't get enough of the poor birds. The poultry industry has become a mass-production ordeal. It begins on the farms, where chickens and turkeys are reared under crowded conditions. Infectious bacteria in the fecal material of one bird are easily passed to the next bird. In the processing plant, eviscerating machines can accidentally cut open a bird's intestines, spewing bacteria all over equipment that contaminates count-less birds. Chill baths, where thousands of defeathered chickens are cooled down, also spread infection.

Such conditions produce more contaminated chickens than the government can possibly control. Reagan-era budget cuts reduced the number of meat and poultry inspectors, while automation in processing increased the number of birds to be examined. Although obviously unhealthy-looking birds are rejected, those contaminated with salmonella or campylobacter may look perfectly normal and be passed over.

The Health Effects on Children

The good news is that salmonellosis is rarely fatal. Symptoms appear within eight to forty-eight hours of eating contaminated food and include diarrhea, abdominal pain, fever, nausea, and vomiting. Symptoms generally disappear after several days in otherwise healthy individuals. Fatalities are rare, about 1 percent of affected people, although in children under one year the rate can be as high as 7 percent, in part because their immune system is immature. Another factor predisposing infants to salmonellosis is that the normal, adult complement of protective intestinal bacteria (flora) is not present until about the age of two.

Illness from campylobacter is different. Because it strikes two to five days after contaminated food is eaten, the cause is rarely uncovered. Fewer than one case in a hundred is thought to be reported. Symptoms include watery diarrhea followed by blood-streaked stools and severe abdominal pain, fever, and chills. Children are the most severely affected.

The Safe Handling of Meat

Such illnesses can be prevented by careful handling, cooking, and refrigeration. The tips listed below should be heeded when preparing all meat products—not just poultry—to guard as well against various gastrointestinal diseases caused by Staphylococci, Shigella, and *Clostridium perfringens*, poisoning from *Clostridium botulinum* in canned foods, tapeworm in beef, and trichinosis in pork.

• Prevent cross-contamination by keeping raw meats and their drippings (blood) from touching other food. This applies to meat that has been refrigerated, since some bacteria can live for up to three weeks in the cold.

—Wrap meat in paper or plastic bags or put in separate containers.

—Wash all utensils, cutting boards, plates, and work surfaces that have been in contact with raw meat with hot, soapy water.

—Wood cutting boards are hard to clean. Keep a separate one for raw meat, or better yet, switch to an acrylic board.

—When washing up juices from raw meat, don't use sponges—bacteria survive in them and will contaminate the next item you wipe. Use paper towels and dispose of them immediately.

—Wash hands thoroughly with hot water, soap, and a nail brush after preparing raw meat. Alternatively, wear disposable or washable rubber gloves when handling it.

—Don't put cooked meat on the same plate it sat on before cooking.

• Keep meats refrigerated. This retards the growth of bacteria. Refrigerator temperature should be 40° F or lower, freezer temperature 0° F or lower. Some microbes like campylobacter are killed by freezing temperatures; buy frozen meat, or freeze it several days before using.

• Thaw meat in the refrigerator, the microwave, or while cooking. If you are in a hurry, thaw meat in a watertight bag under cold running water. Never thaw on the countertop, as bacteria multiply quickly, especially on the meat's surface.

• Poultry, beef, and pork should be cooked thoroughly. This includes hamburgers and hotdogs. Poultry should not be consumed if it is pink near the bone. Use a meat thermometer to make sure meats are cooked to the appropriate internal temperature. Insert the tip into the thickest part of the meat, avoiding fat and bone. Reheat leftovers to an internal temperature of 165° F, and let sauces and gravies reach boiling.

• Be vigilant when cooking with a microwave. Microwaves cook unevenly. Try cooking meat at lower temperature for a longer period of time, and make sure it is rotated.

• After cooking meat, refrigerate it until serving time. Spores from some organisms can survive cooking and germinate and grow at room temperature. Gravy, broth, and large pieces of meat should be cooled to refrigerator temperature within two to three hours. Refrigerate the meat in a shallow container, which will allow all parts to cool uniformly. Don't crowd the container; allow room for the circulation of cold air. Cooked poultry can be refrigerated safely for up to four days. Freeze it after this.

• If your recipe says to let meat sit after removing from the oven, do it. This lets steam out, which finishes the cooking process. It is especially important for microwave cooking.

• If you are using a crockpot, be careful of the temperature. Danger arises when meat is held below 140° F for two or more hours. Cooking should be at 200° F for eight to ten hours. Since not all food in the pot reaches this temperature when the thermometer reads 200° F, you should:
  —Run the crockpot on high (300° F) for the first hour, then turn it down to low (200° F).
  —Never use frozen meat.
  —Warm the meat and the water before cooking.

• Once the package is opened, use hotdogs and lunch meats within three to five days. Discard hotdogs in cloudy liquid; bacteria may be present.

• Refrigerate poultry stuffing until you are ready to stuff and cook the bird. Stuffing should be loosely packed for uniform cooking. Check stuffing for doneness (165° F) with a meat thermometer after the bird is removed from the oven.

• Don't eat food from cans that are leaking, bulging, badly dented, or

• Select meat with less fat, and trim excess fat off. Remove the skin of poultry.

• Don't grill frozen meat. The outside will burn before the inside is adequately cooked.

• Precook thick cuts of meat in an oven or microwave to reduce grilling time.

• Wrap meat in foil to protect it from smoke, or place heavy-duty aluminum foil or a drip pan between the meat and the coals to catch dripping fat. You could also purchase a grill in which the coals are positioned vertically, avoiding the problem of fat dripping onto them (see Resources).

• If the grill is smoking badly from dripping fat, reduce the heat, temporarily remove the meat, or increase the distance between coals and meat.

• Scrape off charred parts of the meat before eating.

• Baste with barbecue sauce rather than fat or oil.

• Use "liquid smoke" in sauces and marinades. This can be found in the condiment section of the supermarket. It is made from smoke derived from burning charcoal and then filtered to remove tars and most PAHs.

• Avoid mesquite and other softwood charcoal. They produce too much dangerous smoke. One study showed that hamburgers cooked over mesquite contained forty times more benzo-a-pyrene than those cooked over hardwood coals.

Eggs: Unbroken but Not Untainted

Another casualty of the mass production of poultry is the egg. Beginning in 1985, outbreaks of salmonellosis were traced to uncracked eggs. How were the eggs contaminated? It is now thought that salmonella infect the yolk of an egg as it is being formed in the ovary, before the shell forms. Unfortunately, infected eggs look perfect; fortunately, a few precautions will protect your family:

• Don't eat raw eggs or foods prepared with them, like Caesar's salad, hollandaise sauce, eggnog, and homemade mayonnaise. Restaurants often use pasteurized liquid eggs, which are not widely available to home cooks.

• Thoroughly cook eggs. Boil them for at least seven minutes; poach for five; fry for three on a side; and scramble until firm throughout. For sunny-side-up, eggs should be cooked until the white is completely firm.

• Never allow your children to lick a bowl containing batter made with raw eggs.

spurting liquid when opened, from jars that are cracked or have a loo
bulging lid, or from canned foods that have a foul odor. These foods m
contaminated with the bacterial toxin that causes botulism. They shou
carefully wrapped and disposed of. Don't taste the food, not even a littl

For more information, contact the U.S. Department of Agriculture's
and Poultry Hotline (see Resources).

Is Irradiated Poultry Safe?

The Food and Drug Administration (FDA) says yes. In 1990, it ap
the use of irradiation to control salmonella and other bacteria on
Zapping food with radiation is similar to pasteurizing milk. It doe:
all the bacteria, but enough to retard spoilage. Irradiated poul
requires proper handling and refrigeration by the consumer.

Irradiation damages the DNA of contaminating microbes so th
multiply. The process does not make food radioactive. However, it
chemical changes in food that lead to the formation of so-called r
products. Whether these are harmful to humans is still a matter c
There is also concern that irradiation destroys some of the nutritic
and that it produces radiation-resistant bacteria.

Until there are more answers to the questions about irradia
producers are likely to shun the process. Major food companies li
Oats, H. J. Heinz, and Ralston Purina have publicly announce
not produce or purchase irradiated food products. New York, N
and Maine have banned the sale and manufacture of irradiated f
their borders.

If you want to avoid irradiated products, look for the la
required by the FDA, a flower in a circle accompanied by the wo
by irradiation."

Are Grilled Meats Safe?

Cooking on the barbecue is one of America's favorite summ
Food cooked this way is undeniably delicious. However, the
risks associated with grilling. When meat is cooked over h
drips onto the heat source and emits potentially carcinoger
called polycyclic aromatic hydrocarbons, or PAHs. PAHs
smoke and are deposited back on the meat. They can also fo
meat when it is charred. Benzo-a-pyrene, one of the mos
causes cancer in laboratory animals. However, the risk to
fully understood. To be on the safe side, and especially
regularly eats grilled food, you may want to follow the sug

FISH: IS IT FRESH OR IS IT FOUL?

Fish is gaining in popularity, and the reasons are obvious. Most varieties are low in fat and high in omega-3 fatty acids, which may protect against heart disease and certain cancers. Fish is also high in protein, B vitamins, and important trace minerals. However, there are potential problems with a diet high in fish. The risk of becoming sick from natural and man-made toxins as well as microbes and parasites in seafood is increasing. Take fish from the Great Lakes, for example. Despite some progress over the past three decades cleaning up pollution in the lakes, invisible toxic chemicals from shore-side industries end up in the fish that goes to market.

Don't rely on the government to protect you from contaminated fish. The FDA, which is responsible for inspecting fish, is sadly short of resources. In 1987, it managed to inspect about nine pounds of fish for every million pounds consumed.

If you choose to include a generous portion of fish in your children's diet, you should be aware of the risks and what you can do to minimize them.

Industrial Chemicals and Heavy Metals

These substances contaminate many of our fishing waters, notably near the largest coastal cities: Boston, New York, San Diego, Seattle. More and more fish from these waters are exhibiting cancerous tumors.

Fish are the primary food source of PCBs, industrial chemicals used until 1971, the year they were banned. They are still present in the food cycle. For example, they remain embedded in the bottom of Lake Michigan, where they are slowly released over the years to contaminate more fish. PCBs cause liver cancer in rats and are toxic to the fetuses of some animals. Studies have not yet found a link between PCBs and human cancer, although some studies suggest a link to low birth weight, as well as impaired reflexes and memory in infants born to mothers who eat fish from polluted waterways.

Toxic metals such as lead, cadmium, chromium, arsenic, and methyl mercury from mining, chemical manufacturing, and sewage sludge have also found their way into the fish we eat. Some of these metals accumulate in shellfish, others in large predatory fish like shark, swordfish, tuna, halibut, and marlin. They cause a variety of serious, chronic illnesses in humans, for instance, kidney damage, mental impairment, and cancer. Children are especially vulnerable.

Microbial Contamination

Microbes in raw or undercooked shellfish like oysters, mussels, and clams are a major source of seafood-related gastroenteritis. Although this illness tends to be relatively mild, for children with compromised immune systems, it can be fatal.

The Worm in the Sushi

Infection from nematode worms has been popping up in the West, Alaska, and Hawaii. Although still relatively uncommon, it is on the rise because of the increasing popularity of Japanese-style raw fish. The worms are passed from marine mammals to fish like Pacific red snapper and salmon.

Two major types of infection (called anisakiasis) can occur. One is relatively mild, although the victim can have the decidedly unpleasant experience of coughing up a live worm within a few days of eating raw fish, or feeling a worm in the throat. In the more serious anisakiasis, nematode larvae penetrate the walls of the stomach or intestine, producing abdominal pain, fever, nausea, vomiting, and/or diarrhea, symptoms that are often misdiagnosed. The only real treatment involves surgery. The best way to avoid problems is not to eat raw fish. If you and your family cannot live without sushi, eat it only in reputable Japanese restaurants; the chefs know good fish from bad. (Most cases of anisakiasis occur from fish prepared at home.) To eliminate the risk of parasites from sushi, sashimi, or ceviche prepared at home, serve seafood that has been frozen at least a week.

Natural Environmental Toxins

These toxins, which can contaminate fish, accounted for approximately 19 percent of all reported foodborne disease outbreaks between 1983 and 1987. Fish are thought to become toxic by feeding on microorganisms that inhabit coral reefs. Fish that appear and smell normal at the time of capture may be toxic.

Ciguatera poisoning is the most frequently reported seafood-related disease worldwide and the most common nonbacterial food poisoning in the United States. Illness comes from eating tropical fish like snapper and sea bass. Most cases occur in southeastern Florida and Hawaii, though an increasing number of Americans are contracting the disease in the Caribbean and the South Pacific or from eating fish imported from these areas. Gastrointestinal symptoms of ciguatera poisoning include nausea, diarrhea, and abdominal pain three to five hours after ingestion. Neurological symptoms begin twelve to eighteen hours after eating contaminated fish. Symptoms include tingling and numbness of lips and tongue, as well as temperature reversal (cold objects feel hot). Death is rare, but neurological symptoms can linger for months or years.

The odorless and tasteless ciguatera toxin is not destroyed by cooking, freezing, or salting. You may want to avoid feeding your family susceptible fish or at least follow the Caribbean maxim: Don't eat a fish that is larger than your dinner plate. The larger and older the fish, the more likely it is contaminated with ciguatera toxin.

Scombroid fish poisoning is caused by the toxin, scombrotoxin, found primarily in tuna, bluefish, and dolphinfish (mahi mahi). It is produced when bacteria decompose fish flesh. Scombrotoxin is not destroyed by cooking and can be present without signs of spoilage. Several minutes or hours after being ingested, it produces nausea, diarrhea, headaches, rash, facial flushing, and sweating. To prevent scombrotoxin formation, keep susceptible fish refrigerated at all times or on ice.

Tips for Eating Fish

The following suggestions should help protect your children from the dangers of contaminated fish:

• Limit your children's consumption of freshwater fish and fish caught in coastal waters. These fish include striped bass, bluefish, sockeye salmon, herring, perch, freshwater bass, catfish, carp, lake whitefish, and lake trout. Especially if they come from waters close to major cities, industrial areas, and agricultural lands, these fish are likely to be contaminated with toxic chemicals, pesticides, and heavy metals. Catfish and carp are the most toxic; they feed near the bottom of lakes, where they are exposed to toxic chemicals in sedimentation.

The least contaminated fish are those like pollack, haddock, and cod that inhabit offshore waters (more than three miles out). These fish are also low in fat.

• Limit consumption of fatty fish like herring and salmon. Fat-soluble chemicals concentrate more heavily in these fish.

• Discard the skin, belly flap, dark meat, and internal organs before cooking. These tissues store chemicals.

• Choose young (smaller) fish for consumption. They have had less time to build up contaminants. The National Wildlife Federation estimates that you have a 1 in 10,000 chance of getting cancer if you eat just eleven meals of 30-inch-long Lake Michigan trout. Thirty meals of 20- to 30-inch trout yield the same risk of cancer.

• If you are pregnant, be extra careful of the source of your fish. Avoid Great Lakes fish. Limit your children's consumption of Great Lakes fish, especially those from Lakes Michigan and Ontario.

• If you go fishing, check with local authorities to see if they have issued health advisories against fishing in the area.

• Purchase seafood only from licensed markets. Be especially wary of raw shellfish. If in doubt, ask the seller to show you the certified shipper tag that comes on shellfish.

• All seafood should be kept cold at all times to prevent bacterial growth. Refrigerate at 32° F, in the coldest part of your refrigerator (lower shelves at back). Refrigerate cooked seafood and use it within three days. Thaw frozen fish in the refrigerator overnight, not on the kitchen counter.

• Cook seafood as soon after purchase as possible, preferably the same day. The internal temperature should reach 140° F. Fish is cooked when the translucent flesh becomes opaque and flakes with a fork. Do not eat raw shellfish.

• Handle raw and cooked seafood separately to prevent bacterial cross-contamination. For the safe handling of raw fish, refer to the tips for the handling of chicken and other meats mentioned earlier in the chapter.

• If your family consumes a lot of seafood, you may want to try surimi. This is deboned and minced fish that is frozen and used to produce seafood products like artificial crab legs. The fish are raised in environmentally controlled waters.

MILK: STILL WHITE BUT NOT SO PURE

Milk is synonymous with purity and wholesomeness. It's the complete food. "It does the body good," says a TV commercial aimed at children, the most avid milk drinkers next to teenage boys. However, two miracles of modern science appear to be threatening our milk supply. These are a genetically engineered hormone and drugs like antibiotics and sulfas. How great is the threat?

Bovine Growth Hormone

Bovine growth hormone (BGH), made by the pituitary gland of cows, allows them to produce milk. By means of gene cloning, scientists have removed from cows the DNA that codes for BGH and put it in bacteria that produce large quantities of pure hormone. Bacterial-made BGH, which differs only slightly in structure from the natural hormone, can then be sold to farmers who inject it into their cows to boost milk production up to 25 percent. The FDA says the milk is safe, as do manufacturers and many scientists. But there are scientists and organizations that do not agree.

What is the controversy about? One scientist, Samuel Epstein of the University of Illinois, thinks synthetic BGH may indirectly stimulate premature growth in infants and breast cancer in women. There are those who prophesize economic ruin. The small dairy farmers fear that the resulting increase in the milk supply will drive down prices and make their farms unprofitable. There are also the pragmatists that don't want their good name associated with a controversial product; it is simply not good for business. These include some of the nation's largest supermarket chains, which have at least temporarily refused to sell dairy products from cows experimentally treated with BGH.

The FDA is expected to approve the use of BGH sometime in late 1991. In fact, in an unprecedented move the agency published a summary of thirty years of scientific studies on BGH for review by the public (see Resources). (Such reports are normally published after a drug is approved by the agency.) The review concluded that genetically engineered BGH presents no increased health risk to consumers. Specifically, the concentration of BGH in cow's milk does not increase significantly as a result of the treatment of cows with synthetic BGH. In any case, most BGH is destroyed by pasteurization. Even if humans were to ingest traces of synthetic BGH, it would most likely be inactive in the human body. An independent panel appointed by the National Institutes of Health (NIH) at the request of Congress to review the data also concluded that synthetic BGH is safe for use in cows.

Despite these conclusions, controversy lingers. One major concern is that these groups came to their conclusions using incomplete data; the companies that make the hormone won't release all the data from their studies. The missing data, according to Epstein, is critical. Further dimming prospects for the swift approval of BGH, the highly respected Consumers Union, publisher of *Consumer Reports* magazine, concluded after its own analysis of the data that the FDA should reopen its evaluation of the health effects of BGH.

Until the critics are quieted, BGH-treated milk is not likely to be embraced by the consumer. Milk from cows treated with the hormone will most likely be labeled as such, so consumers can make an easy choice. Meanwhile, BGH milk promises to be a hot political issue.

Milk and Meat from Medicated Cows

In early 1990, independent studies by the Center for Science in the Public Interest, the *Wall Street Journal*, and the FDA revealed traces in milk of various antibiotics and sulfa drugs. These drugs are routinely fed to livestock to control disease and make the animals grow faster. This widespread practice may be harmful to the consumer for several reasons.

First, traces of drugs like penicillin, tetracycline, and sulfas, consumed unknowingly in milk, may cause an allergic reaction in sensitive individuals. Milk is the most common cause of food-related allergy. It is possible that in some cases the allergen may not be a constituent of milk but a drug given to the cow.

Second, some drugs detected in trace amounts in milk are suspected human carcinogens. Sulfamethazine, a sulfa drug given to cows and pigs to speed growth and prevent respiratory disease, has been detected in milk and pork. Eating these foods may slightly increase our risk of cancer. The FDA intends to ban this drug. Nitrofurazone, another suspected carcinogen, is widely used by farmers to treat milk cows. An FDA proposal made over a decade ago to ban this drug has not been realized. The FDA suspects that farmers are covertly using at least thirty unapproved drugs whose health effects are not known. FDA tests are not capable of detecting most such drugs.

Third, massive use of antibiotics in animals is making these drugs less effective for treating humans. When animals are given antibiotics continuously, the bacteria they harbor become resistant. When we consume food products from drug-treated animals, we are ingesting antibiotic-resistant bacteria. Some of these bacteria, like salmonella or campylobacter, can make us ill. If doctors prescribe antibiotics that the bacteria are resistant to, the treatment will be ineffective. This happened in Michigan. Eighteen people became ill and one died from a strain of salmonella that was resistant to antibiotics used in animals. To make matters worse, if you ingest animal products containing antibiotic-resistant bacteria, resistance can spread to other potential disease-causing bacteria in your intestine. Sixty percent of beef cattle, 90 percent of pigs, 90 percent of calves for veal, and almost all poultry are fed antibiotics.

What Can Parents Do?

There is no need to eliminate milk or other animal products from your children's diet. Depriving your children of the nutrients in these foods just to avoid trace amounts of drugs would be a mistake. You can, however, look for meat and milk that come from drug-free animals. Consult with your grocer or local natural-food store. Alternatively, Americans for Safe Food (see Resources) can provide you with a list of mail-order beef, pork, and poultry producers that don't use growth stimulants or pesticides.

Also, it is prudent not to overuse antibiotics or pressure your pediatrician to prescribe them for viral infections for which they will have no effect. When your child is on antibiotics for a bacterial infection, administer the entire prescription; otherwise you will be encouraging the growth of antibiotic-resistant bacteria. If your child is undergoing antibiotic treatment for an infection, he or she will be more susceptible to antibiotic-resistant, disease-causing bacteria like salmonella. Don't feed your children raw or undercooked meat. Lastly, ask your senators and representatives to support legislation to prohibit the use of medically important antibiotics as well as cancer-causing drugs in animals.

RESOURCES

Products and Services

Hermelin, 130 McCormick Ave., Suite 109, Costa Mesa, CA 92626. Telephone: (800) 545-4808.

 Sells grills for barbecuing foods in a vertical position.

Sources of Information

Americans for Safe Food, 1875 Connecticut Ave. NW, Suite 300, Washington, D.C. 20009. Telephone: (202) 332-9110.

 Write for a list of mail-order suppliers of organic and growth-stimulant-free beef, pork, and poultry.

"Biotechnology and Milk: Benefit or Threat?" Michael Hansen, 1990.

 This report addresses health and safety issues of BGH use in milk. Available for $5 from Consumer Policy Institute, 101 Truman Ave., Yonkers, NY 10703. Telephone: (914) 378-2000.

"Bovine Growth Hormone: Human Food Safety Evaluation." Judith C. Juskevich and C. Greg Guyer. *Science Magazine* 249 (1990): 875–84.

Eating Well: The Magazine of Food and Health

 A beautifully illustrated food magazine that features timely articles on safety issues. Call (800) 344-3350 for subscription information.

Food and Drug Administration, 5600 Fishers Lane, Rockville, MD 20857. Telephone: (301) 443-3170.

 Write to the Consumer Affairs Office (HFE-88) for "Food Safety," a general information source. The Office of Public Affairs (HFI-40) will send you "For Oyster and Clam Lovers the Water Must Be Clean," which contains information on eating shellfish safely.

Food Marketing Institute, 1750 K St. NW, Washington, D.C. 20006.
> Will send you "The Food Keeper," a brochure on proper refrigerator and freezer storage, pantry storage, and foods that need special handling. Send 50¢ with a legal-sized, self-addressed, stamped envelope.

Safe Food—Eating Wisely in a Risky World. Michael Jacobson, Lisa Lefferts, and Anne Witte Garland. Los Angeles: Living Planet Press, 1991.

Seafood Safety. Farid E. Ahmed (ed). Washington, D.C.: National Academy Press, 1991.

U.S. Department of Agriculture, USDA-FSIS, Rm. 1163-S, Washington, D.C. 20250. Meat and poultry hotline: (800) 535-4555 (Washington D.C. residents 447-3333).
> Has the helpful booklets "A Quick Consumer Guide to Safe Food Handling," "Talking About Turkey," and "Meat and Poultry Safety: Questions and Answers About Chemical Residues." For answers to questions on meat and poultry, call the hotline.

The University of Illinois, Office of Agricultural Communications and Education, 69G Mumford Hall, 1301 West Gregory Drive, Urbana, Il, 61801. Send for the *Complete Guide to Home Canning* (AIB539) ($9.00).

CHAPTER 11

Subtracting Additives from Processed Foods

The Problem. The subject of food additives like monosodium gluta-mate, aspartame, and nitrite is highly controversial within the scientific community. Yet the government has not banned these additives from foods, many of which are consumed in great quantity by children.

The Risk to Kids. Some additives can cause acute adverse reactions in sensitive individuals. Others may cause long-term effects like cancer and brain damage.

What to Do. After learning the potential health risks of additives like monosodium glutamate, parents can make informed decisions about whether to limit or eliminate additive-containing foods in their children's diet.

As every parent knows, it can be a real challenge to get children to eat healthy foods. The sheer number of food products on the market can overwhelm even the most conscientious parent trying to determine what is healthy. And before choices are made, television and acquaintances have already strongly influenced the child's diet. How difficult it is to keep those beautifully packaged, delicious, heavily processed foods from the mouths of children!

The countless items jammed onto America's grocery shelves contain a vast array of chemical additives now available to the food producer. These have a variety of functions. Preservatives like sodium nitrite inhibit spoilage that causes discoloring and lost flavor. Agents like vitamins and minerals improve the nutritional value of foods. Coloring agents like FD and C Blue No. 1 make them attractive. Flavoring, both natural and synthetic, is the most common additive. And then there are texturizing agents like mono- and diglycerides, added to change the "mouthfeel" of food.

HOW IS THE GOVERNMENT PROTECTING US?

In 1958, the Food Additives Amendment was passed. This, along with the Color Additive Amendment of 1960, is the most recent change to the Federal Food, Drug, and Cosmetic Act adopted in 1938 to protect our nation's food supply. These amendments turned the tables on food manufacturers, who were adding just about anything to food. The amendments required that an additive be proven safe *before* it entered the food supply. In addition, both amendments contained the Delaney Clause banning any additive found to induce cancer in man or animal (see page 89 for further discussion of the Delaney Clause).

GRAS Foods

Along with the Food Additives Amendment, Congress established the so-called GRAS (generally recognized as safe) list. GRAS substances had never been rigorously tested in animals; they were considered safe by qualified scientists based on a history of use in foods with no apparent ill effects on human health. Most of them could be added to food in any concentration. GRAS additives included salt, cornstarch, mustard, licorice, and garlic, as well as vitamin A and benzoic acid.

The rationale for the GRAS list was to focus scarce testing resources on new additives for which little historical safety data was available. At the time, however, health information on many of the listed substances was scarce. Just because a chemical made the original GRAS list did not guarantee its safety. Cyclamate was one. The artificial sweetener, later shown to cause bladder cancer in animals, was banned in 1970. After this incident, President Nixon directed the FDA to reevaluate the safety of all the GRAS chemicals for use as food additives. The reevaluation of GRAS substances, conducted by an expert advisory group, continues to this day.

Chemicals created since 1958 can be added to the GRAS list if the manufacturer can show enough data from human experience or scientific literature to prove that they are safe.

Additives not on the GRAS list must be tested for their health effects on animals. Manufacturers perform these tests before petitioning the FDA for use of the additive in foods. If low doses of an additive cause harmful physiological effects or internal organ damage in animals, or if any dose causes cancer (Delaney Clause), the chemical should be banned from use. However, additives are not always banned when they are supposed to be. Saccharin is a case in point. This sweetener came on the market in 1879. It was shown to cause cancer in lab animals in 1977. The FDA applied the "zero-risk" standard of the Delaney Clause and announced it would ban the product. But a large public outcry led by special-interest groups erupted, which resulted in the Senate overturning the ban. The risks to health were outweighed by the benefits of saccharin in diet control. While saccharin is still on the market, its popularity has diminished with the advent of another artificial sweetener, aspartame.

Some thirty additives have been banned because the health risks of eating them were deemed greater than the benefits. Others that have been demonstrated to be risky in man, and some that cause cancer in animals, have not been banned. These include saccharin and the additives discussed below. Their continued use in our food supply is controversial.

WHAT ARE THE HEALTH CONCERNS FOR CHILDREN?

The concern over additives is twofold. First, they have made possible the proliferation of junk foods, which may be eaten instead of more nutritious fare. Second is the intrinsic health risk of additives. Acute health effects are rare because few additives are present in food at high enough levels to cause an immediate reaction like poisoning. The major problem is hypersensitivity in certain individuals and chronic, long-term diseases like cancer. There is no direct evidence linking consumption of additives with chronic health problems like cancer; animals studies have provided most of the information on such risks.

FOOD ADDITIVES

The following additives are approved for use by the FDA and are currently in our food supply. You may want to restrict foods in your children's diet containing them because their safety is still debated.

Monosodium Glutamate
Monosodium glutamate (MSG) is a sodium salt of glutamic acid, one of the twenty amino acids. MSG was designated GRAS based on a long history of

no apparent side effects. It has been used as a flavor enhancer in Asia since antiquity. MSG is used to intensify the taste of many foods, including meat, condiments, soups, and pickles. It is thought to be responsible for the "Chinese restaurant syndrome" experienced by sensitive individuals shortly following a Chinese meal. Symptoms include everything from headaches and numbness in the neck to nausea and vomiting. MSG can trigger asthma attacks and heart palpitations.

Food processors must list MSG on the label if it is added directly to the product. They do not have to disclose MSG if it is added as part of "hydrolyzed vegetable protein" (HVP) or "natural flavoring." Hundreds of products contain HVP, of which up to 20 percent can be MSG. Some products even claim to be MSG-free when they contain HVP.

What do scientists say about all this? For twenty years they have been debating the safety of MSG. A major concern is that glutamate functions in the brain as an excitatory neurotransmitter, that is, when released by some nerve cells, it stimulates other nerve cells to higher levels of activity. In excess, glutamate can stimulate neurons to death. Dr. John Olney, a neurophysiologist at Washington University, feels that this poses a special risk to young children. His studies on rodents and monkeys show that MSG can damage the hypothalamus. Animals whose brains were damaged by glutamate in infancy exhibit stunted growth, obesity, and reproductive problems in adulthood. Infant animals also proved more susceptible than adults to brain damage from MSG. Dr. Olney feels that human infants could be similarly affected and that the damage may not be noticeable until years later, making it difficult to trace the cause.

These findings, along with prodding from Olney, convinced baby food manufacturers to remove MSG from their products. However, MSG is still found in countless foods consumed by young children, among them canned soup and frozen pizza. Olney argues that a 20-pound child can receive an almost toxic dose of MSG from eating a 6-ounce serving of certain brands of instant soup. Not all neuroscientists agree, and the controversy rages on.

Aspartame

Aspartame was approved as a tabletop sweetener in 1981, and in 1983 for use in diet soft drinks, the most common source for children. It is marketed under the brand names Equal and Nutrasweet. There are lingering questions about how the FDA's approval process for aspartame was conducted. Questions about such effects as tumors seen in some tests have not been adequately investigated.

Dr. Olney argues that children may be at risk because aspartame contains aspartate, an amino acid that activates the same receptor in the brain that glutamate does. Aspartate's excitatory properties can also kill neurons. Infant animals fed aspartate develop the same symptoms that glutamate causes, stunted growth and obesity. Although there is no consensus among scientists about the safety of aspartame, these facts alone should alert parents with young children; this additive is among the most popular and ubiquitous in our food supply.

The acceptable daily intake for aspartame in children has been estimated to be 4 to 6 milligrams per pound of body weight. If a 44-pound child consumes just one 12-ounce can of soda containing 200 milligrams of aspartame, he or she will have reached that limit.

There is another health problem about which there is no controversy, and it relates to the other amino acid component of aspartame—phenylalanine. One out of every 20,000 babies has the genetic disease called phenyketonuria (PKU). A baby with PKU cannot metabolize phenylalanine; if it rises to toxic levels, the result can be mental retardation. All newborns are tested for PKU, and affected babies are put on a special phenylalanine-free diet. They must be protected from aspartame for the rest of their life. Parents have to be vigilant when taking their PKU child to restaurants, which, unlike producers of packaged foods, are not required to advertise warnings about food. Another concern involves pregnant women who carry the PKU gene (1 percent of the population), that is, those who do not have PKU themselves but may be carrying a fetus with the disease. In their diet, these mothers may unwittingly expose their fetus to phenylalanine, possibly causing mental retardation. In 1984, the American Academy of Pediatrics' Committee on Genetics and Environmental Hazards expressed concern about high levels of aspartame consumption by pregnant women and young children.

Nitrite

Approximately 7 percent of our food supply is treated with this additive. You have seen it listed on packages of bacon, hotdogs, luncheon meat, cured meat, and fermented sausage. It is also used in some fish and poultry products. Nitrite, mainly in the form of sodium nitrite, is put in foods to kill the microbes that cause botulism. Nitrite also enhances the flavor of meat and keeps it from turning brown or gray.

The controversy over nitrite involves its tendency to react with other chemical components of foods during cooking or in the stomach to form nitrosamine. The latter, shown to cause cancer in many animal studies, is one of the most potent of known carcinogens. Studies in humans have suggested a connection between a high level of nitrite and cancer of the stomach and esophagus. (The Japanese consume a large amount of nitrite and have the world's highest incidence of stomach cancer.) Although provocative, these studies do not prove that nitrite causes cancer in humans. Industry reminds us that it is crucial to preventing botulism, a serious food poisoning.

But is nitrite crucial? There are nitrite-free turkey and chicken lunch meats on the market that are perfectly safe, suggesting that the botulism argument may be overblown. Brands include Weaver Turkey Breast, Louis Rich Honey Turkey Breast, Oscar Mayer Roasted Chicken Breast, Mr. Turkey Chicken Breast, and Eckrich Lite Roasted Turkey Breast. In response to concerns over nitrite, bacon manufacturers have added either sodium ascorbate (vitamin C) or vitamin E–related substances to their products to slow the formation of nitrosamine.

If your children like bacon, you can minimize the risk by cooking it in a microwave. Some studies have shown that the lower cooking temperature of microwaves prevents the formation of nitrosamine. If you fry bacon, throw out the grease because it contains a high level of nitrosamine, and be sure to ventilate the cooking area, as nitrosamine is volatile. Have your children drink a vitamin C beverage like orange juice when eating nitrite-containing meats, or better yet, choose brands of lunch meat without nitrite.

Sulfites

Sulfites, which include sodium bisulfite and sulfur dioxide, have a variety of functions. They prevent the browning of potatoes and of canned and frozen vegetables, prevent the darkening of dried fruits like raisins and apricots, and control the growth of undesirable microbes in soft drinks, fruit juice, beer, wine, sausage, fresh shrimp, clams, crabs, and pickles.

Five to 10 percent of asthmatics are sensitive to sulfites, as are some nonasthmatics. Reactions include nausea, diarrhea, acute asthma attack, loss of consciousness, and shock. Some twenty deaths have been associated with sulfite, the only food additive known to have been a direct cause of death. Sulfites cause genetic mutations in test microorganisms, although they have not yet been shown to do so in animals. The FDA has banned the use of sulfites on fresh fruit and vegetables. Sulfites must be listed on wine and packaged food labels when they exceed 10 parts per million.

Red Dye No. 3

This is the stuff that adds a bright red color to cherries in fruit cocktail, pistachio shells, cheese wax, toothpaste, candy, and lipstick. In 1982, food industry studies showed that this colorant caused thyroid cancer in male rats. Application of the Delaney Clause would have banned it. But it wasn't banned, at least not until 1990, and then only partially. Further scientific data from the National Institutes of Health suggested that the dye might enter brain tissue and interfere with neuron transmission. This phenomenon could lead to behavioral or learning problems in children, among other things.

The fruit cocktail industry waged a war to prevent the banning of the dye, and it was kept on the market until January of 1990, when at least 20 percent of its applications were banned. It can no longer be used in any cosmetic, including lipstick, powder, blush, and shampoo. It is also banned from use in drug capsules, coughdrops, cake frosting, chewing gum, and wax on cheese. The FDA says it will eventually outlaw No. 3 in food products like fruit cocktail, pistachios, candy, fruit juices, breakfast cereals, and Jello. Nobody knows when. To be on the safe side, limit your children's consumption of foods containing this as well other artificial food coloring—Yellow No. 6, Blue No. 1, and Blue No. 2, for which data on carcinogenicity is equivocal.

Fat Substitutes

Fat substitutes are the newest food additives. In the spring of 1990, NutraSweet won FDA approval to market Simple Pleasures, a frozen dessert containing Simplesse, a fat substitute made of proteins from egg whites and milk. Simplesse replaces part or all of the fat, thus decreasing total calories. It may eventually be approved for use in a wide array of other products, including mayonnaise, salad dressing, butter, and cheese spread. Trailblazer, a similar product from Kraft General Foods, is expected to be approved soon. And Procter and Gamble has a heat-resistant product called Olestra that is awaiting FDA approval.

Although fat substitutes look like miracles that will allow us to satisfy our craving for fat without the calories, many scientists and nutritionists are worried. A major concern is that people may eat more junk food containing these products than healthy food like fruit and vegetables. This would be especially detrimental to growing children. Also, it is predicted that the craving for fat will still need to be satisfied—that people will eat high-fat food in addition to fat-substitute products. (The introduction of artificial sweeteners saw a concomitant rise in sugar consumption per capita.)

Fortunately, fat substitutes aren't the only game in town, and parents have alternatives. If you want your children to have low-fat desserts, look for certain nonfat yogurts, "lite" frozen tofu, sherbet, and frozen fruit bars. They are as low or lower in fat than Simple Pleasures.

FOOD UNDER WRAPS:
ARE PACKAGING MATERIALS DANGEROUS?

Several packaging materials on the market leak chemicals into food when heated to high temperatures. Just how dangerous these chemicals are to our health is not yet known.

Heat-Susceptor Packaging

Microwave ovens don't brown and crisp food the way a conventional oven does. Food-packaging manufacturers have a solution to this problem called heat-susceptor packaging. With heat-susceptor packaging, you get crisp french fries and pizza crust. The packaging is not difficult to recognize. Found in nearly a third of the microwaveable foods sold, it usually appears as a strip or disk of metallized plastic. This material absorbs microwaves and heats up quickly so that food sitting on top of it browns and stays crispy.

The heat susceptor is usually a metallic film placed over plastic called polyethylene terephthalate (PET), which is laminated to paperboard with adhesive. PET has been approved by the FDA but not for use at the high temperatures that a heat susceptor can reach in a microwave oven. In 1988, the FDA tested various types of heat-susceptor packaging by heating corn oil in them. Every package released chemicals, including PET, into the oil. The adhesive that holds this type of packaging together, composed partly of trace amounts of suspected or known carcinogens, also migrates into cooked foods.

The FDA has asked packaging manufacturers to identify which chemicals are migrating and how much, and to investigate possible health hazards. It is suspected that the FDA will not ban heat-susceptor packaging but rather will list acceptable chemicals to guide industry. In the meantime, it would be wise to avoid microwaveable products in heat-susceptor packaging—pizza, fries, waffles, popcorn, and breaded fish—until their safety has been better evaluated.

Microwaveable Containers

Be careful of microwaving foods in any old plastic container you have at home. Don't grab the leftover margarine or yogurt tubs for use in the microwave. These containers, as well as some marked "microwave safe," may actually be leaking chemicals into your food. The FDA, although concerned, is not yet regulating what goes into them. So until more is known about their safety, beware. Glass and Corning Ware are still the best containers for microwaving.

Dual-Oven Trays
There are many products on the market whose special packaging allows them to be cooked in both microwaves and conventional ovens. Dual-oven packaging releases chemicals like PET when used in a conventional oven at temperatures of 350–400° F, but not at a microwave oven's lower temperatures. To be on the safe side, cook foods in dual-oven trays only in the microwave; transfer the food to other cookware if you put it in a conventional oven.

Cling Wrap
In 1987, the safety of plastic "cling" wrap that contains vinyl chloride polymers (PVC) was called into question by a British study. The study showed that a chemical called DEHA, commonly used in PVC food wrap as a plasticizer or softening agent, migrates into fatty food during microwave cooking, at room temperature, and even in the refrigerator. The highest levels of DEHA, a possible animal carcinogen (it causes cancer in mice but not in rats), were found in fatty foods like pork that came into contact with the wrap during microwaving. The lowest levels were found in nonfatty food like vegetables and fruit, and in food not actually touching the wrap during cooking.

In the United States, DEHA is a common component of PVC wraps used by supermarkets for meats, fish, and cheese. The chemical probably contaminates the surface of these foods. We don't know whether PVC cling wrap bought for home use contains DEHA; this is information the manufacturer does not have to divulge.

If possible, order your meats, fish, and cheese at a counter where it can be wrapped fresh, preferably in waxed or brown paper. At home, keep fatty foods like meat out of direct contact with cling wrap, even in the refrigerator.

Lead in Cans
Another problem with packaging is the presence of lead in cans. This toxic metal is dangerous to your children even at low levels (see chapter 1 for a discussion of lead poisoning in children). Lead is found in the solder used to seal some cans. Solder leaks lead into acidic food (fruits and tomato sauce) and juice as well as into less obvious food like tuna. A freshly opened can of juice can contain lead above the maximum level set by the EPA. And after

five days of open storage, during which oxygen increases corrosion, the lead level becomes dangerously high.

Surprisingly, there is no FDA ban on lead in food cans. There is, however, a "voluntary reduction program" that was worked out between industry and the FDA in the mid-1970s. Lead solder was removed from the seams of baby food and formula cans, and can manufacturers agreed to replace lead-soldered cans gradually. This resulted in a dramatic reduction (about 75 percent) of lead solder in cans. Unfortunately, lead cans do remain on the market. Some manufacturers continue to use old equipment, others use the new method of jet-soldering, which reduces but doesn't eliminate lead. And there is no standard governing imported cans.

There are a number of ways to protect your children from food with lead in it. First, don't buy lead-soldered cans. You can identify them by tearing off the label; if you see a crimped joint (edges folded over) smeared with silver-gray solder, that is lead. Lead-free cans have no side seam or else have a thin, sharply defined, blue-black line along the welded seam. Avoid jet-soldered cans, which have a line down the side, slightly wider than the blue-black line of a lead-free can, and a crimp or two. Be sure to inspect imported cans, and take advantage of the brands that are starting to label their cans lead-free. Also, never leave food in opened cans; store food in nonmetallic containers. Unopened cans containing high-acid foods like tomato and grapefruit juice, applesauce, mixed fruit, and vinegar-based sauces should be stored no longer than twelve months.

High-acid food also causes tin to leach from cans, especially when stored after opening. There are concerns about the effect of tin on growth and bone formation as well as blood-cell count and the immune function. You should be concerned about tin if your children consume a lot of canned food.

Lead in Pottery

Don and Fran Wallace of Seattle, Washington, happened to bring home Italian coffee cups purchased during their travels abroad. They used the cups regularly and scrubbed them clean with abrasive cleanser. Then they developed serious, puzzling symptoms that their doctors failed to recognize as high-level lead poisoning. Not until they were near death was it discovered that they had been poisoned by the lead leaching from their cups. After recovery, Don got a degree in public health and set up his own laboratory and mail-order company to test for lead in pottery. The Wallaces are left with the question of what long-term health problems they will face as a result of lead poisoning.

Potters add lead to paint and glazes to achieve brighter hues. If pottery is fired at temperatures too low to stabilize the outer lead glaze, it may leach lead into food. The leaching process is encouraged by highly acidic food like coffee, vinaigrette, orange juice, and wine. Scrubbing with an abrasive cleanser promotes deterioration of the glaze.

The FDA set limits on lead release in ceramic dinnerware in 1971 and tightened them in 1980. These force U.S. producers to observe proper glazing techniques, and their products should be safe. However, concern lingers over ceramic products made by unregulated foreign producers and made in the United States before 1970. Approximately 60 percent of all dinnerware sold in the United States is imported. The job of inspecting foreign imports is difficult because of a large volume. A random survey in 1987 showed that 4.4 percent of all ceramic products violated FDA lead guidelines.

The U.S. Potters Association believes that most European and Japanese wares are safe but that many products from developing nations may not be. In the mid-1980s, FDA violations occurred primarily in products from Asian nations, especially China. A certification program initiated in that country in 1987 should increase the safety of its dinnerware. Mexico is a problem because many of its ceramics are made by small manufacturers not subject to quality control. Be careful when traveling to Mexico or other foreign countries, as your family may be eating from pottery that leaches lead. And don't buy dishes for eating if they are made by home craftsmen or studio potters.

How can you tell if your dishes are leaking lead? You can test ceramics and lead-soldered cans with two mail-order lead-testing kits (see Resources). Both give quick results and are easy to use and relatively inexpensive, according to *Consumer Reports* magazine. However, neither kit is sensitive to low levels of lead, which can be dangerous to your children.

Other Sources of Lead

Another source of leaching lead is found on drinking glasses decorated with leaded enamel logos. Many of these are distributed by gas stations. Although the logo is on the outside of the glass, toxicologists note that this is the area toddlers often lick.

Who would have thought that the printing and designs on most plastic bread bags would contain lead? Well, they do—scientists in New Jersey found that 87 percent of the wrappers they tested had lead and other toxic metals like chromium. These ingredients can be hazardous if bags are turned inside out and used to pack your child's lunch.

Another potential hazard is lead-crystal baby bottles, which have become a popular baby gift. In 1991 researchers at Columbia University demonstrated that after sitting just 15 minutes in a lead-crystal bottle, warm formula absorbed almost 200 micrograms of lead per liter. This is considerably higher than the EPA's current standard for lead in drinking water. Lead also has been shown to leach from crystal decanters and wine glasses into the beverages they are holding. Obviously, the prudent course is never to give your children liquid from such containers.

RESOURCES

Products and Services

Frandon Enterprises, P.O. Box 300321, Seattle, WA 98103.

Sells the Frandon Lead Alert Kit.

Lead Check, P.O. Box 1210, Framingham, MA 01701. Telephone: (800) 262-LEAD.

Markets Leadcheck Swabs, used to test for lead in ceramics and soldered cans.

Sources of Information

"Amino Acids: How Much Excitement Is Too Much?" Marcia Barinaga. *Science Magazine* 247 (1990) 20–22.

Explains the controversy over the health effects of glutamate and aspartame.

Center for Science in the Public Interest, 1875 Connecticut Ave. NW, Suite 300, Washington, D.C. 20009. Telephone: (202) 332-9110.

Has a handy directory to over fifty food additives and their health effects called "The Eating Smart Additives Guide" (no. 50, $3.95). Members receive *The Nutrition Action Health Letter,* which features articles on additives.

The Complete Eater's Digest and Nutrition Scoreboard. Michael F. Jacobson. Garden City, New York: Anchor Press/Doubleday, 1985.

Half of this book is devoted to food additives: their history, their health effects, and the controversy about their use.

A Consumer's Dictionary of Food Additives. Ruth Winter. New York: Crown Publishers, 1989.

"Lead Astray: The Poisoning of America." Michael Weisskopf. *Discover* (December 1987): 68–77.

An interesting article that reviews the subject of lead.

National Organization Mobilized to Stop Glutamate, P.O. Box 367, Santa Fe, NM 87504.

Provides information on the health effects of MSG.

CHAPTER 12

Curbing Contaminants
in Water

The Problem. Despite progress in reducing water contamination by sewage treatment and curbs on industrial discharge, a significant amount of America's water is polluted with toxic and carcinogenic chemicals, many of which are released by unidentified and uncontrolled sources.

The Risk to Kids. Children are especially vulnerable to water pollution because, relative to body size, they drink more liquid and absorb more pollutants than adults. Their developing organs make children more vulnerable to poisoning by certain pollutants like lead and nitrates.

What to Do. Parents can ensure the safety of the family drinking water by learning about the source of that water, its likely contaminants, how and where to test it, and appropriate purification strategies and equipment. Parents can also educate their children about water protection through programs like Adopt-a-Stream.

Drinking water safety has become a major concern to Americans regardless of where they live or what the source of their water is. While public health and environmental experts say most American tap water is safe, problems do occur. Microbes and a growing list of chemical contaminants are finding their way into surface water and in groundwater, which collects many feet below the earth's surface. Conventional disinfection and piped delivery systems can introduce additional contaminants. Children have high fluid requirements—babies the highest of all—which means they are significantly exposed to these pollutants.

Take, for example, the problem that surfaced in 1985 at Fernald, Ohio. Radioactive waste buried by a rural nuclear weapons plant under contract with the U.S. Department of Energy leached into local drinking water wells. Some residents had assumed the facility, named the Feed Materials Production Center, made pet food; and many were unaware when the well water uranium levels rose to thirty times the level considered safe. One area family spent five years unknowingly drinking the contaminated water and feeding it to their baby boy. Now they must deal with the fact that exposure to radioactive materials can result in cancer after a lag period of up to several decades.

Or take the simple process of piping water into homes. This can introduce lead into drinking water. The still-maturing nervous systems of babies are especially vulnerable. In 1985, in an affluent neighborhood of Washington, D.C., parents of a ten-month-old baby noticed she was acting cranky, this on top of chronic digestion problems. A blood test given to all city children revealed she had high lead levels. Subsequently, the drinking water in her home was found to have three times the level of lead considered safe. The city advised residents of her neighborhood to drink bottled water only.

SURFACE WATER CONTAMINANTS

Surface water, which includes lakes, rivers, and reservoirs, supplies about half of America's drinking water. Rivers and lakes have been severely polluted by wastes discharged from adjacent industries and from municipal wastewater treatment plants. The 1972 Clean Water Act required states to make rivers and lakes clean enough for fishing and swimming, and it tightened limits on "point" sources of pollution, that is, those that could be identified and located. The Clean Water Act and the 1974 Safe Drinking Water Act improved sewage treatment for 42 million Americans. Some problems remain, such as noncompliance by point-source polluters. They are legally bound to remove toxic pollutants from their discharges, but not all do. Other industries continue to release unregulated poisons into water. Pulp mills that bleach paper with chlorine discharge dioxins, which the EPA considers probable carcinogens.

Non-Point-Source Reduction

After progress was made in cleaning up point-source pollution of surface water, the problem of non-point-source pollution emerged. This is pollution by sources not easily identified or controlled, such as urban runoff, which is a mixture of rainwater and contaminants like soot and spilled motor oil, and rural runoff, which contains pesticides and fertilizers washed from farms, golf courses, and yards. Fertilizers in surface waters reduce the oxygen available to fish, and pesticides poison aquatic life and enter the food chain. Other non-point sources are faulty septic tanks that leak untreated sewage, and landfills or chemical storage facilities that release toxic chemicals.

The EPA and the states agree that non-point sources account for half of all water pollution. Amendments passed in 1987 to the Clean Water Act address the non-point source problem, for example by requiring states to develop management programs that reduce urban and agricultural runoff through physical barriers, street cleaning, and reduced use of farm chemicals.

Contaminated Sediments

Not enough attention has been paid to the contaminated sediments on the bottom of surface water. This "toxic grime" is the final resting place for pollutants like heavy metals and organic chemicals. In 1987 an EPA study concluded that nearly every major U.S. harbor and body of surface water had sediment contamination. One problem is disposing of the silt left over from dredging for navigation. The most contaminated silt is placed in combined-disposal facilities (CDFs) in upland sites or holding ponds. CDFs are rarely fenced off, covered, or posted. In Chicago, children swim in a CDF sitting next to a public park. The EPA has no national criteria that restrict pollution of underwater sediments or trigger their cleanup and protection. Recently, a coalition of groups wrote the "Citizen Charter on Contaminated Sediments" asking Congress to establish and fund a program to address these needs.

GROUNDWATER CONTAMINANTS

Half of America depends on groundwater for their drinking needs. Because it sits under layers of filtering soil and semiporous rock, groundwater generally has fewer kinds of organic contaminants than surface water, but what contaminants it does have may be more highly concentrated. Once groundwater is contaminated, the process of pumping it to the surface, cleaning it, and returning it to the aquifer is difficult and expensive.

Over 200 different chemicals have been detected in groundwater. More than 60 pesticides have contaminated wells in 30 states. Much groundwater contamination is from non-point sources—millions of underground gas and chemical storage tanks, as well as the runoff from city streets, industry, agriculture, and mining operations. Regional groundwater contamination by pesticides began to hit the news in the 1970s. Dibromochloropropane (DBCP) was found in California, followed by aldicarb in Long Island, New York, and ethylene dibromide (EDB) in Florida. The EPA reported in 1990 that 10 percent of community drinking water wells and 4 percent of rural domestic wells contained at least some pesticides. The EPA has produced a pamphlet "Pesticides in Drinking Water Wells" on this problem (see Resources).

About 70 percent of America's 400,000 landfills lack linings to prevent the leaching of hazardous chemicals. Bacteria from overburdened or faulty septic systems leak into groundwater. Other sources of pollution include highway salt and de-icing compounds, runoff from barns and parking lots, and some 19,000 hazardous waste sites.

WATER TREATMENT

Both surface and groundwater are commonly treated by filtration and chlorine disinfection. First, water is pumped into a holding tank where a chemical such as chlorine or alum (aluminum sulfate) is added to help sediment clump together and to kill microbes. Next, the water sits in a sedimentation basin while solid particles sink to the bottom. The water is strained to remove larger debris. Then it is sent through a final filter of gravel and sand to remove more dirt and microbes. Chlorine is added again for disinfection. Some treatment systems add other materials to improve water taste or hardness. The latter is considered a healthy property: People living in geographical areas with hard water have a longer life expectancy. Sometimes fluoride is added to prevent tooth decay.

Filtration and chlorination remove miscellaneous debris and control pathogenic bacteria such as those that cause typhoid fever and cholera. Some purification systems apply chlorine three times during water treatment. Chlorine is less effective in killing water-borne protozoa (single-celled organisms) and viruses. One resistant protozoan is *Giardia lamblia*, an intestinal parasite. *Giardia* can be a problem in surface water systems that chorinate but don't filter their water. There is no EPA standard for protozoa or viruses in water. The EPA feels they are best controlled by full water treatment, meaning both chlorination and filtration.

The Hazard of Chlorine

Chlorine reacts with natural organic debris like leaves in surface water to form undesirable byproducts such as trihalomethanes (THMs), which have been linked with an increased risk of cancer and birth defects. THMs are found in low concentrations in treated surface water throughout the country. The EPA has set a total THM limit of 100 parts per billion (ppb) in drinking water serving populations over 10,000. The standard doesn't apply to smaller systems that are often dependent solely on chlorination. If the THM level in your water exceeds 100ppb, you can reduce it with activated carbon filters (see below). Treatment plants can reduce THM levels by adding chlorine to water when it cannot encounter plant debris. To reduce the taste and smell of chlorine, keep water for several hours in an uncovered pitcher or agitate water in a blender for several minutes.

Aluminum

Adding alum during water treatment adds some aluminum to drinking water. This raises questions about the safety of lifetime exposure to dietary aluminum. An unusual amount of aluminum has been found in the brain tissue of Alzheimer's Disease patients. A recent study found that people drinking aluminum levels of .11 parts per million (ppm) or more faced a 50 percent greater risk of getting Alzheimer's Disease than when the level was lower than .01 ppm. The EPA has an unenforceable "secondary standard" for aluminum in drinking water, which is a range of .05 to .20 ppm. Infant exposure begins early because both breast milk and formula can contain significant amounts of aluminum.

Fluoride

People living in areas where water contains naturally occurring fluoride experience less tooth decay. Fluoride has been added to drinking water at the rate of about 1 ppm since 1945. It is now in over half the nation's drinking water. Fluoride can cause troubling side effects in children. Some teeth can become mottled with brown or yellow spots when fluoride reaches levels of 1 ppm; mottling may grow severe at 4 ppm. Levels of 3 ppm can cause bone thickening and the formation of bony outgrowths. Research studies have established a weak link between fluoride and bone cancer in male rats, but the data is of marginal statistical significance. There is considerable debate over the EPA standard for fluoride, which is 4 ppm. This high limit helps states with naturally occurring high levels avoid expensive defluoridation. The EPA reviews fluoridation safety data every three years. Recent studies to reevaluate fluoride's benefits are hampered by the relative lack of fluoride-free populations. Even people drinking non-fluoridated water may use fluoridated toothpaste or eat processed foods containing fluoridated water. Fluoride can be removed from water using a reverse-osmosis system.

WHY ARE KIDS MORE VULNERABLE
TO WATER POLLUTION?

Children consume more water relative to body weight than adults. Adults consume 2 to 4 percent of their body weight in fluids, healthy infants 10 to 15 percent and children under five years 4 to 8 percent. Children also absorb some toxic inorganic chemicals like lead at a higher rate than adults.

Children are especially vulnerable to toxic water contaminants like nitrate and lead because their organ systems are not yet mature. They are also more vulnerable to carcinogenic and radioactive contaminants.

Nitrate Poisoning

Dave Dolaz of Mason County, Illinois, who raises mostly popcorn on his 1,000-acre farm, mixes a liquid fertilizer directly into his irrigation system. He has to drive five miles to town to haul drinking water back for his family, which includes two youngsters, because his well water is contaminated with nitrates. Nitrates are a common groundwater pollutant resulting from chemical fertilizer. Their mature digestive systems protect adults from nitrate damage, but in very young children nitrates are converted to nitrites, which then interfere with the ability of blood to carry oxygen, a condition called methemoglobinemia. Poisoned babies develop blue lips and skin and in severe cases may experience brain damage and death. An EPA nationwide well-water survey found in 1990 that more than half of sampled wells were contaminated by nitrates.

Lead Poisoning

Lead is a heavy metal that is very poisonous to the nervous system. It has received a lot of attention because of recent findings that it is not safe even at comparatively low levels. Low levels may impair fetal mental development, reduce birth weight, cause learning disabilities and hyperactivity in children. The EPA estimates that one in six Americans drinks tap water containing excessive lead levels. The metal usually enters drinking water from lead plumbing in water mains, service-connector pipes, and homes. Homes built before 1930 are the most likely to have lead plumbing. Homes built before the 1986 ban on lead solder may have copper pipes with lead solder or flux. Solder and flux can leak lead for about five years. Some chrome-plated faucets may contain lead. In hard-water areas, minerals like magnesium and calcium tend to coat the inside of pipes after about five years, effectively stopping lead from entering water. Soft water, which has fewer dissolved minerals and is more acidic, promotes pipe corrosion and thus the escape of lead into water. Children can also be exposed to lead through water coolers and fountains, which may contain lead-lined water tanks and lead soldering. Lead levels forty times as high as the EPA's recommended standard have been measured in water from coolers. School districts are supposed to test their water for lead levels, but this has not been strictly enforced by the EPA.

Carcinogenic Organic Chemicals

A number of carcinogenic organic chemicals are found in water: the THMs mentioned earlier, solvents like trichloroethylene (TCE), pesticides, and industrial chemicals like polychlorinated biphenyls (PCBs). PCBs are now banned, but they remain in surface-water sediments. Over 90 percent of the U.S. population has a measurable level of PCBs in the blood, and PCBs have been found in significant quantities in breast milk.

TCE is one of many widely used industrial solvents that have contaminated the well water of rural households and towns. In 1979, TCE and several other organic chemicals from toxic waste dumps forced the closing of several wells in Woburn, Massachusetts, after a cluster of childhood leukemia was discovered. Well water was also linked to an increased incidence of other childhood diseases and birth defects. The incidence of all these health problems declined after the contaminated wells were closed.

Small organic molecules like chloroform are usually volatile, meaning they evaporate easily, especially when water is heated or agitated. While showering or washing, children absorb volatile organic chemicals in significant amounts through the skin or the lungs.

Radioactive Contaminants

Groundwater contamination by nuclear weapons plants has occurred at seven locations nationwide. For example, over 100 square miles of groundwater has soaked up billions of gallons of radioactive waste from the Hanford Reservation in Richland, Washington.

Radon, the naturally occurring radioactive gas that seeps into homes from the ground, can contaminate well water. It evaporates from water that is heated or agitated during showering, dishwashing, and laundering. (See more on radon gas in chapter 2, "Ridding Your Home of Radon.") As many as 8 million Americans may have excessively high radon levels in their tap water. The EPA estimates that approximately 100 to 1,800 people die annually from lung cancer caused by inhaling radon released from household water. The water most likely to contain radon is from private wells and community systems serving less than 500 people. This water doesn't get aerated as it would in a larger water system, where radon evaporates before reaching household taps.

Radium and uranium, which occur naturally in the earth, can enter groundwater. For example, radium is a problem in the groundwater from deep aquifers in some far western suburbs of Chicago. These communities are trying to reduce their exposure by switching to more shallow wells. The EPA standard for radium in drinking water is 5 picocuries per liter, although the EPA proposed in 1991 to relax the standard to 40 picocuries. Call your state environmental agency or radiation control office to check whether there is a known problem in your area's groundwater.

TESTING YOUR DRINKING WATER

How healthy your tap water is depends on where it comes from, what might be contaminating it, how it is treated, and how carefully it is monitored. The EPA sets standards, called maximum contaminant levels (MCLs), for widespread or particularly toxic contaminants. Currently it regulates thirty-six contaminants, which are listed in the EPA pamphlet "Is Your Drinking Water Safe?" (see Resources) and in the Code of Federal Regulation (CFR). The CFR is updated annually and distributed to libraries nationwide. (The EPA plans to regulate over sixty contaminants by July 1992 and over eighty by the spring of 1993.) You can also call the EPA's Safe Drinking Water hotline (see Resources) for the most recent list of regulated contaminants.

Who Does the Testing?

About 60 percent of Americans get their water from a public supplier that is legally required to check for bacteria, trace metals, and other contaminants. You can request a copy of the test results. The 1986 amendments to the Safe Drinking Water Act give citizens the right to ask the following questions:

- What is the source of my water?
- How is my water purified?
- What contaminants have been tested for?
- What are past and present contaminants?
- What contaminant levels violate current federal drinking-water standards?
- How is the public notified of violations?

These questions and more are listed in a "citizen questionnaire" in Ralph Nader's *Drinking Water Newsletter* (see Resources). Data from your water supplier won't reveal lead contamination. Fortunately, it is not expensive to run a test for lead (see below). Some cities test residents' water for free.

Big cities like St. Louis run hundreds of water analyses every day. Other systems may neglect mandatory tests, can't afford expensive monitoring, or aren't required to test for dangerous chemicals and toxic metals. You may have to look to other sources, such as the state EPA, or run some tests yourself.

The rest of the American population, 40 percent, uses private well water. Although this water is not routinely tested by a supplier, the state EPA may have run tests on water in your area. Or the local public health department may know if your groundwater is polluted. Some local governments test well water free or for a small charge. Your county health department may test wells if there is a specific problem like gasoline odor or discoloration.

Where Tests Are Done

Your state EPA can probably supply a list of certified testing labs. You can also try a national mail-order testing service (see Resources). A testing service will send you a kit with collection bottles and instructions; you mail the water sample by overnight delivery and receive the results, along with help in interpreting them, within a few weeks. Prices vary, as does performance reliability. If results show significant contamination, have a second test run by another lab before proceeding. Prices at WaterTest, a national mail-order lab, begin at about $30 for a lead test and rise to $195 for a more thorough test of organic chemicals. Some labs offer an all-purpose test for $89, which covers bacteria, common organic chemicals, and dangerous metals and minerals.

There have been reports of scams in which an "inspector" claiming to be from the water utility, a university extension service, or an environmental group pressures people into buying an expensive and inappropriate water purification system. If you are approached by a suspicious salesperson, call the EPA Safe Drinking Water hotline. This hotline is also a good source of information about certified testing labs.

How to Choose a Test

Because it is expensive to run tests for all water contaminants, you must try to select the proper tests. The staff of *Consumer Reports* has prepared the following guidelines for well-water users:

- Test for organic chemicals if you live within two miles of a gas station, refinery, chemical plant, landfill, or military base.
- Test for nitrates and pesticides if you live in an agricultural area.
- Test for lead if your house is over thirty years old or if you have lead solder in your water pipe connection.
- Test periodically for bacteria, inorganic compounds, and radon no matter where you live.

Be sure to test your water if you notice a change in its appearance, taste, or smell.

Because surface water is easily contaminated, drinking water taken from it is usually treated and tested. If you drink from untreated, private surface water, you should test for the same things a public supplier does: bacterial contamination and any federally or state-regulated contaminants. Consider also what particular contaminants may be present as a result of land use in your area. The state EPA should be able to give information on this. Don't forget to test for lead from plumbing and delivery systems.

WHAT TO DO ABOUT CONTAMINANTS

Several steps can be taken if tests reveal your water is significantly contaminated. First, you may want to switch to bottled water, at least temporarily. Labels can be confusing. Spring water isn't necessarily from a spring, and some bottled water is from the tap. Still water is noncarbonated, sparkling water is carbonated, mineral water has high levels of minerals. Look for a bottler who will reveal the source of the water, and can provide a water analysis done by an accredited laboratory. Look for a brand low in sodium and disinfected by ozone. Check to see if the water was rated in a test of fifty brands run by *Consumer Reports* (January 1987). This test revealed that one brand contained organic chemicals. People with no water contamination should remember that bottled water is not always superior to tap water. A 1991 FDA survey of bottled water revealed that 31 percent of 52 U.S. brands tested were tainted with bacteria, although not fecal bacteria. One brand contained *Pseudomonas aeruginosa* bacteria, which some experts regard as a threat to infants.

Next, identify the source of the contaminant. Start with yourself if you have a private well. Have you disposed of any toxic wastes in your yard? Then consider the land use of the watershed area no matter whether your water source is private or public. For example, check your vicinity for dumps, industrial plants, buried gas tanks, and agricultural runoff. If your lead levels are high, check your home and water-delivery system for lead plumbing. Alert the public health department about the problem.

Finally, you should consider using a home purification unit to remove contaminants. Keep in mind, however, that the EPA doesn't regard home units, which are point-of-use (POU) devices, as a permanent solution to water contamination when a community treatment system can be upgraded. POU devices are hard to maintain and to monitor.

WATER PURIFICATION SYSTEMS

Purification systems can cost over $500, installation fees over $200—expensive, but probably cheaper than the long-term purchase of bottled water. Avoid inexpensive devices that wear out quickly or filters that require frequent replacement. Look for systems that are practical to use, like those that fit under the counter. If your water does not contain radon gas or volatile organic chemicals, which can be inhaled or skin-absorbed, you will only need to treat the water you drink. The average family needs about three gallons of purified drinking water a day. It is important to realize that no one device can remove all contaminants. You can select the correct system(s) only after finding out exactly what contaminants your water has. There are three main water purification systems: activated carbon, distillation, and reverse osmosis.

The Activated-Carbon System

This system uses carbon filters to trap volatile organic chemicals, chlorine, THMs, and pesticide residues. These filters improve the taste and odor of water. They don't remove toxic lead or nitrates. The filters must be changed regularly or they may provide a breeding ground for bacteria. The systems are priced at $150 and up, but keep in mind the cost of replacing filters. You can extend the lifetime of a carbon filter by running water first through a sediment filter.

Activated carbon filters available for shower faucets (see Resources) reduce the volatile organic chemicals that can be absorbed through the skin or the lungs. These filters vary in price, and because of the volume of water used in the shower, users may have to change them regularly. The other way to reduce volatile gases is to ventilate. Crack a window or operate an exhaust fan while children bathe or shower.

The Distillation System

This unit heats tap water until it turns to steam, then captures the steam and condenses it into distilled water. It is very effective at removing toxic metals and dissolved solids from water, but not small organic molecules like chloroform. Distillation units consume a lot of energy and produce a lot of heat, which is uncomfortable in the summertime. The boiling flask accumulates mineral scale if hard water is used, and it must be regularly cleaned to operate efficiently. Most units cost over $300.

The Reverse-Osmosis System

This forces tap water through a semipermeable membrane that holds back most kinds of contaminants (toxic metals, fluoride, nitrates, radium, and lead) but not all organics. The membrane will not remove chloroform. When reverse osmosis is coupled with an activated-carbon filter, almost all contaminants are removed. A good under-the-counter reverse osmosis unit runs from $450 to $850, with maintenance costing $25 to $100 a year.

Bacterial contaminants can be removed to some extent by distillation or reverse osmosis. Another recommended method is the ultraviolet (UV) device, which uses intense UV radiation to kill bacteria. Water may have to be prefiltered before UV treatment, and not many devices are available for home use.

Where to Purchase a Unit

It is important to purchase your unit from a reliable dealer, preferably one that belongs to the Water Quality Association in Lisle, Illinois (see Resources). This group sets standards and handles consumer grievances. The dealer you choose should be able to answer questions about how a system works, what contaminants it will remove, what the installation and maintenance costs will be, what difficulties may arise, any problems covered by warranty, and whether you can replace filters yourself. It is also a good idea to consult the January 1990 issue of *Consumer Reports* for a rating of brand-name units.

The National Sanitation Foundation (NSF) certifies purification devices for specific functions such as the removal of lead or THMs. You can get a list of certified models by writing the NSF (see Resources). When you purchase a unit, the company's sales literature must disclose which contaminants the unit is certified to remove.

SPECIAL STRATEGIES TO REDUCE CONTAMINANTS

Sometimes a unique water purification device or household strategy works best to get rid of a contaminant. This can be illustrated by current methods of treating lead, nitrate, and radon contamination.

Strategies for Lead

In 1991 the EPA changed its 50 parts per billion (ppb) lead limit in drinking water to an action level of 15 ppb, which applies to lead levels at home faucets. If more than 10 percent of homes in a public water system exceed the limit, the water authority must take action to control corrosion or replace lead delivery pipes. Critics point out that children will remain at risk because the new rules don't limit lead contamination in the water supply itself. Also, utilities have up to twenty-one years to comply. Thus families should still check out their own lead levels at home. Take action if your level is over 15 ppb, and if there are children or pregnant women in your home. Drink bottled water or use distillation or reverse osmosis to remove the lead. If lead levels are low, and you wish to keep them that way, use cold water for cooking and drinking, especially when making baby formula, because hot water leaches more lead out of pipes. If you have lead pipes, let the tap water run for at least a minute first thing in the morning.

Strategies for Nitrates

Well water in agricultural areas should be tested regularly for nitrates and treated if levels exceed the EPA's safe limit of 10 ppm. Distillation or reverse-osmosis units will do the job. Another way to avoid nitrates, which are most common in shallow wells less than 50 feet deep, is to dig down to an uncontaminated water source. But this is not a permanent solution, since deeper water may someday become contaminated.

Strategies for Radon

In 1991 the EPA proposed a standard—effective at the earliest in 1993—of 300 picocuries of radon per liter for drinking water. If you think you have a problem, read the EPA booklet "Removal of Radon from Household Water" (see Resources), which discusses when and how to test for radon in water, and how to determine if your water is contributing a significant amount of radon to indoor air. If you decide to take action, your first option is to ventilate the rooms where radon can escape from water: the bathroom, the laundryroom, and the kitchen. If you choose to treat your water, you should treat all water entering the house. Two useful devices are the granular activated-carbon (GAC) unit and the home aerator. The GAC unit costs between $800 and $1,200 installed. It should reduce radon levels by about 90 percent. Place the GAC unit outside of the home or behind shielding indoors. The EPA describes the proper use and maintenance of GAC units in the pamphlet "Removal of Radon from Household Water." The home aerator is even more expensive at $2,500, but it may be more efficient.

Home aerator tanks are placed in the basement, where they pump air through water, causing radon to bubble off and move through a pipe for venting outdoors.

HOW TO PROTECT WATER

Water can be cleaned up and protected against contamination. Specific grass-roots strategies exist for treating both surface water and groundwater. But remember: Cleaning up begins at home, as does protecting your children.

You can make a difference by not purchasing hazardous household chemicals and not pouring any you do have down the drain. Families should look for nontoxic alternative products and dispose of any hazardous substances through community waste-collection programs. These and related activities are explored in detail in part 5, "The Healthy Cleanup."

Well owners can take special steps to protect groundwater. They should check their property for pollution sources like leaking septic or fuel tanks and agricultural chemicals. Septic tanks become overburdened when additions are built on homes. Maintain your septic tank by having it professionally pumped out every three to five years. Any chemical poured down the drain and into a septic system can eventually reach groundwater. Never use an organic chemical like TCE to clean your septic system. Underground gas or heating-fuel tanks are designed to last no more than several decades. If you use tanks, rule out the possibility of leakage by keeping a record of their contents or by testing them with a tank dipstick. If your tanks are no longer in use, make sure they have been drained. See if you can reduce your use of pesticides and chemical fertilizers. Practice integrated pest management (IPM) and use organic fertilizers when possible. These simple and safer practices are discussed in chapter 5. Don't mix pesticides or clean pesticide equipment near your well. Make sure abandoned wells are properly sealed with clay material. Ask for details from your state EPA.

Teaching Your Children

It is important to teach your children about protecting water. Let them help you in your efforts, or get them involved in their own project. Many children in Washington State have participated in an innovative program called Adopt-a-Stream, for which a guidebook is available from the University of Washington Press (see Resources). This program helps clean up polluted streams and restore wildlife. One example is the work done by students and teachers at Pigeon Creek, in Everett. Students cleaned debris out of the stream, tracked down sources of pollution in the water recharge area, and distributed pamphlets on how to properly dispose of hazardous household waste. They put signs on storm drains, Outlet to Stream, Dump No Waste, and began releasing baby salmon into the creek. Three years later, shiny adult salmon returned after a twenty-five-year absence. Children who participate in the twenty-year-old "Save Our Streams" program from the Izaak Walton League (see Resources) can measure stream pollution by analyzing water bugs. They can use the League's "Citizen Directory for Water Quality Abuses" to report oil slicks or fish kills.

RESOURCES

Products and Services

All Seasons Marketing, 3576 Wesley Chapel Rd., Marietta, GA 30062. Telephone: (404) 565-2397.

> Sells a test kit to identify lead in soldered joints of copper plumbing.

Baubiologie Hardware, 200 Palo Colorado Canyon Rd., Carmel, CA 93923. Telephone: (800) 441-8971 or (408) 625-4007.

> Sells the Rainshow'r, a shower faucet that conserves water and reduces chlorine.

Environmental Hazards Management Institute, 10 Newmarket Rd., P.O. Box 932, Durham, NH 03824. Telephone: (603) 868-1496.

> Send $3.75 for the "Water Sense Wheel," a comprehensive guide for understanding drinking water quality and treatment alternatives.

Izaak Walton League of America, 1401 Wilson Boulevard, Level B, Arlington, VA 22209-2318. Telephone: (703) 528-1818.

> Will send information on the Save Our Streams (SOS) Program. Offers a $6 "SOS Kit" to start your own program, and for $1 the "Citizen's Directory for Water Quality Abuses." Will sponsor the Third National Volunteer Monitoring Conference in Annapolis, MD, in the spring of 1992.

NEEDS (National Ecological and Environmental Delivery System), 527 Charles Ave. 12-A, Syracuse, New York 13209. Telephone: (800) 634-1380.

> Offers the ShowerClean water filter, which reduces chlorine levels in water.

University of Washington Press, P.O. Box 50096, Seattle, WA 98145-5096. Telephone: (206) 543-8870.

> Will send you *Adopting A Stream: A Northwest Handbook,* for $11.45.

Water-testing services by mail order. They are listed by region:

> New England and national: WaterTest Corporation of America. Telephone: (800) 426-8378.

> Midatlantic and national: Suburban Water Testing Laboratories. Telephone: in Pennsylvania (215) 929-3666, elsewhere (800) 433-6595.

Midwest and national: National Testing Laboratories. Telephone: in Ohio
(216) 449-2525, elsewhere (800) 458-3330.
West: Water Analysis and Consulting. Telephone: (800) 426-8378.

Sources of Information

American Groundwater Trust Information Hotline. Telephone: (800) 423-7748.
Offers taped information on water quality, water wells, and water protection.
Callers can leave messages. Operated by the National Well Water Association
of Dublin, OH.

Drinking Water Newsletter, P.O. Box 19367, Washington, D.C. 20036.
To order this, send a self-addressed, stamped envelope plus a check or money
order for $1.

EPA. Safe Drinking Water hotline: (800) 426-4791, in Alaska and Washington,
D.C. (202) 382-5533.
Ask how to contact drinking water specialists at your state EPA (see appendix). Request the pamphlet "Is Your Drinking Water Safe?", which lists
federally regulated contaminants, "Removal of Radon from Household Water," and "Pesticides in Drinking Water Wells."

Is Your Water Safe to Drink? Raymond Gabler. New York, Consumers Union: 1988.
Detailed and informative discussion of drinking water hazards and how
individuals and communities can make water safe.

National Sanitation Foundation, 3475 Plymouth Road, P.O. Box 1468, Ann
Arbor, MI 48106. Telephone (313) 769-8010.
Write for a free list of certified water treatment units.

"Protecting Your Rural Well." *Country Journal* 16 (September-October 1989):
23–31.
Good information and advice for well owners, and information on state
wellhead protection programs.

Public Information Office, Water Pollution Control Federation, 601 Wythe St.,
Alexandria, VA 22314-1994. Telephone: (703) 684-2400.
Send a self-addressed, stamped envelope for the brochures "Groundwater"
and "Nature's Way: How Waste Water Treatment Works for You." Also sells
for $49 a "Water Environment Curriculum" (grades 5 to 9) consisting of four
videos, twenty student guides, and one teacher's guide.

Water Quality Association, 4151 Naperville Rd., Lisle, IL 60532. Telephone:
(708) 369-1600.
Provides a list of recommended dealers of water purification units.

Part 4

The Healthy Nap

Hazards in the home and the yard are most commonly associated with those times children are awake and physically active. But there are hazards lurking at the time children seem most protected, when they are resting. To draw attention to their constant, silent presence in the home, this section discusses several such phenomena: chemicals that outgas from synthetic building materials and furnishings, gases from ordinary sources of fuel combustion, microorganisms in household air, and finally, extremely low-frequency electromagnetic fields from conventional electric wiring.

CHAPTER 13

Reducing Harmful Gases

The Problem. Indoor air frequently contains irritating, poisonous, or carcinogenic pollutants from combustion appliances, construction materials, home furnishings, and smoking.

The Risk to Kids. Children, who relative to body size breathe in more air and thus more pollution than adults, spend a lot of time inside. Their developing bodies are more sensitive to respiratory irritants and cancer-causing agents.

What to Do. There are several strategies for reducing dangerous gases, for instance, carefully operating and maintaining appliances, sealing off formaldehyde-containing particleboard, increasing ventilation, reducing indoor humidity, and banning smoking.

No scene is more tranquil and angelic: children safely tucked in their beds. But are they really safe? Consider the case of the Farlie family. Barbara Farlie, an energetic mother and businesswoman, came down with the worst case of flu she had ever had. A dreary but predictable routine set in. After running her son to school, she would return home and go straight to bed, experiencing head pain, fatigue, dizziness, and dimmed vision. Her two sons came down with the same flu-like symptoms. One complained of headaches two hours after arriving home and would fall asleep over his school books. Curiously, the boys' symptoms diminished when they left the house. Barbara finally realized what was wrong after her husband read an article in the newspaper. It was the grim story of a New Jersey family whose pajama-clad bodies were found in front of the TV set, the victims of carbon monoxide (CO) leaking from a faulty heating system. The Farlies' furnace, it turned out, was cracked and leaking CO. Luckily, they found out in time.

Most combustion gases don't kill as fast as CO. Instead, they contribute to long-term health problems like bronchitis, emphysema, and asthma, as well as heart disease and cancer. Where do these gases come from? Normal household activities—smoking, cooking, and heating your home.

Other dangerous gases are released to household air from passive sources like construction material and home furnishings. After their home was insulated with urea-formaldehyde foam, the Rutherford family of Pasadena, Texas, began to experience subtle but debilitating symptoms. They became nervous, irritable, and itchy, and began falling asleep at strange times. The Zenzens of Houston, Texas, had a similar experience. They lived in a three-bedroom mobile home loaded with formaldehyde-containing particleboard. They moved to a new house with less formaldehyde in the construction materials, but they remain hypersensitive to this toxic gas even when exposed to low concentrations.

Formaldehyde is only one of the many potentially toxic vapors that swirl inside our homes. The air of an average American home contains over 350 organic chemicals, some of which cause mutations in test bacteria or cancer in animals or man. The household sources are varied: textiles, carpets, wood products, plastics, copying machines, cleaners, maintenance products, and pesticides.

POLLUTANTS FROM COMBUSTION APPLIANCES

The main pollutants are CO, nitrogen dioxide (NO_2), sulfur dioxide (SO_2), and particulates. They originate from burning fossil fuels in kerosene and gas space heaters, gas stoves, wood stoves, fireplaces, and central heating systems.

CO is an invisible, odorless gas produced when there is an inadequate supply of air for combustion. CO reduces the capacity of blood to carry oxygen to the heart and other organs, resulting in fatigue in healthy people and chest pain and irregular heartbeat in people with heart disease. Higher concentrations can cause headaches, dizziness, weakness, nausea, disorientation, unconsciousness, and death.

NO_2 is a strong-smelling gas formed when gas and kerosene combustion appliances (but not electric or wood-burning appliances) are operating. Health effects range from eye, nose, and throat irritation at low exposure to chronic lung disease like emphysema and death at higher exposure.

SO_2 is produced by burning sulfur-containing fuels in unvented kerosene space heaters or wood-burning stoves. SO_2 can aggravate symptoms in individuals with asthma and bronchitis.

Particulates (visible as smoke) are tiny particles formed from incompletely burned wood in wood-burning stoves and fireplaces, and from smoldering cigarettes. These particles can absorb other toxic substances created by combustion, like the carcinogen benzo-a-pyrene, and carry them into the lungs. The particles can remain in the lungs for months, causing respiratory diseases and exposing the lung to cancer-causing agents.

The Special Vulnerability of Children

Young children are the most vulnerable to indoor air pollution because they spend much of their time at home playing on the floor, where the heavier pollutants settle and can be inhaled. Children are also physiologically more vulnerable than adults to air pollutants, as explained in chapter 3, "Controlling Crumbling Asbestos."

Children are more prone to asthma than adults. Approximately one-third of all asthmatics in the United States are under seventeen, with the youngest children being at greatest risk. Asthma in childhood often leads to chronic asthma and emphysema in adulthood. Polluted air can trigger asthma attacks, adding yet another burden to an asthmatic child's weakened pulmonary system. There is some evidence that certain indoor pollutants may actually cause asthma in young children.

The type and severity of symptoms depend on an individual child's sensitivity. Many are similar to cold and flu symptoms. Most diminish or disappear when the person leaves home. For this reason, it is important to pay attention to the time and place symptoms occur.

Reducing Pollution from Space Heaters

Gas, kerosene, and oil space heaters are commonly used to heat individual rooms. When these appliances are operated without venting, a common practice in the interest of energy efficiency, pollutants such as CO, NO_2, formaldehyde, and SO_2 circulate indoors. After just one hour of operation, NO_2 and SO_2 levels emitted from heaters can exceed the government's outdoor ambient-air standards.

Here are some tips for minimizing pollution from space heaters:

• Don't use unvented kerosene or gas space heaters except for emergency heat, and then only for a short time.

• Follow manufacturer's operating instructions carefully.

• Use the proper fuel in kerosene heaters, not a lesser grade that is likely to contain more pollutants.

• Use space heaters in well-ventilated rooms (crack a window) or large rooms opened up to the rest of the house. You can place an unvented heater in your fireplace, leaving the fireplace damper open to vent combustion gases.

• Never leave a child alone in a room with a space heater.

Reducing Pollution from Wood-Burning Stoves

The energy crisis stemming from the Arab oil embargo of 1973 made the wood stove a popular source of home heating. Interest was rekindled following Iraq's invasion of Kuwait in 1990. Unfortunately, burning wood produces much more pollution per unit of heat than burning oil or natural gas. Air pollutants like CO are released when the wood does not burn completely, for example when air inlets are choked down to produce low heat for overnight use. Most wood-burning pollutants go up the chimney as smoke. However, if a stove is not properly installed, if there are cracks in the stovepipe, or if downdrafts occur, pollutants may stay inside. Measurements done in typical homes with either a wood-burning stove or a fireplace detected levels of pollutants, including the carcinogen benzo-a-pyrene, up to eight times the outdoor levels in the most heavily industrialized cities. Wood stoves are more of a concern than fireplaces because the latter are less frequently used.

In 1988, the EPA set new rules for manufacturers of wood stoves, forcing sharp reductions in emissions of pollutants. When the federal regulations take full effect in 1992, the manufacture and sale of stoves that violate these standards will be prohibited.

Many new stoves use catalytic converters to burn fuel more completely; others employ secondary combustion chambers for this purpose. The newer models consume only half the wood of older models and emit up to 90 percent fewer particulates.

All wood stoves that have been certified by the EPA carry a label saying so. EPA's certification program ensures that all certified stoves (nearly all on the market today) burn cleanly under laboratory test conditions. However, how cleanly a stove burns in your home will depend on how well you operate and maintain it.

Here are some tips for minimizing pollution from both old and new wood-burning stoves:

• Avoid antique or used stoves.

• Choose an EPA-certified stove that is the proper size (in terms of heat output) for the space to be heated. A stove either too large or too small will not be energy efficient and may create more pollution. Consult a heating contractor or your wood-stove dealer for advice.

• Proper installation is critical. It is best to have a professional do it, and it is a good idea to have a periodic inspection to ensure that the vents are working. Catalytic converters on newer models require regular inspection and replacement. Consult your owner's manual for guidance.

• Make sure the chimney is tall enough to vent gases up and away from your home.

• Stove doors should be tight fitting and stovepipe connections secure. To check for cracks or leaks, make a fire and then shut off the supply of air. The fire should go right out. If it continues to burn, there is a leak that needs repair. If door gaskets in older models need replacing, be careful. Gaskets often contain asbestos (older models) and fiberglass (newer models); they should be carefully removed so as not to contaminate indoor air.

• Gas can leak indoors when the stove door is opened to add wood. To prevent this, first open the damper all the way, increasing the draft up the chimney, then open the door.

• Burn dry, seasoned wood for more efficient combustion and less pollution. Hardwoods like oak and maple are preferred to softer woods like pine and spruce.

• Remember that little or no smoke should exit from your chimney if your stove is burning cleanly. The darker the smoke, the more pollutants it contains. If fuel is burning efficiently and cleanly, only white steam should be visible from your chimney.

Reducing Pollution from Gas Stoves

About 60 percent of American families cook with gas stoves. Many households also use the gas stove as a supplemental source of heat. Homes with gas stoves have elevated indoor levels of CO and NO_2. Pollution levels are lowest when combustion is efficient, and combustion is efficient when the right mix of natural gas and air is fed to the flame. The more burners used, the higher the emissions. In one study, the level of CO emitted from a new gas stove was measured in a small house. After just 35 minutes, the peak concentration of CO was 24 ppm (the outdoor concentration was 1 ppm). Almost 3 hours later, the concentration throughout the house was 10 ppm. This exceeds the national ambient-air standard of 9 ppm (averaged over 8 hours) for outdoor air.

Studies of the health effects on children of gas cooking in the home have either been inconclusive or shown a weak association with respiratory illness.

Here are some tips for minimizing pollution from gas stoves:

• Use a range hood over the stove to vent gases from cooking. Otherwise, install an exhaust fan in a nearby window and use it while cooking. This will substantially reduce levels of NO_2 and CO.

• When buying a new gas stove or range, opt for pilotless ignition rather than a conventional pilot light. Pilotless ignition cuts both fuel consumption and pollutants by a third.

• If you have a pilot light, make sure it is well tuned. The flame should burn blue. If the flame burns yellow, pollutants are forming. Make sure the gas jets are unblocked.

Reducing Pollution from Modern Heating Systems

Modern heating systems that are installed and maintained properly pose no health threat. However, leaking, blocked, or damaged chimneys or flues, as well as cracked heat exchangers in furnaces, can release harmful gases and, rarely, fatal concentrations of CO. Periodic and careful inspection of your furnace, chimneys, and flues is critical, and the filters on your furnace should be regularly changed during the cold season. Look for cracks or open joints in the ductwork, and obstructions like nests and soot or yard debris in the chimney. The flue that vents the furnace should be used only for that purpose. Don't vent other appliances like a wood stove through this flue; it could become blocked, preventing the release from your home of dangerous gases.

It is a good idea to install a gas detector with an alarm (see Resources) that will alert you to the presence of CO.

THE PERILS OF PASSIVE SMOKING

By now, everyone has got the message that smoking is dangerous to the health. Recent research also suggests that cigarettes smoked in the house harm children. (About 60 percent of American children live in homes with at least one smoker.) A report issued by the EPA in June 1990 concluded that passive smoking, that is, inhaling the smoke from other people's cigarettes, causes some 3,800 lung cancer deaths in the United States every year. As a result of this finding, the EPA will most likely classify passive or secondhand smoke as a class A carcinogen—a substance known to cause cancer in humans.

Smoking contributes over 4,700 pollutants to the air, including CO, nicotine, formaldehyde, and benzo-a-pyrene. At least 40 are known carcinogens. Sidestream smoke, that rising directly from a burning cigarette, accounts for most of the pollutants released by a lit cigarette.

The Effects on Children

Not even the fetus in the mother's womb escapes the effects of cigarette smoke, considered the most lethal pollutant to which it can be exposed. Smoking mothers have more miscarriages and stillbirths than nonsmokers, and they give birth to twice as many low-birthweight babies (under $5\frac{1}{2}$ pounds), who later suffer the physical, intellectual, and behavioral consequences.

Studies of infants under the age of one show that they are more likely to develop pneumonia and bronchitis as well as allergic diseases such as asthma if their parents smoke. Up to 34 percent of asthma in children may be attributable to maternal smoking. Infants of smokers also show a higher incidence of middle-ear blockage and ear infections, which require hospitalization. After the age of two, the incidence of these infections goes down in infants. In older children of parents who smoke, chronic coughing, wheezing, and phlegm production appear to be more common.

Cancer is also associated with passive smoking. A disturbing 1990 study in the *New England Journal of Medicine* showed that about 17 percent of lung cancer among nonsmokers can be attributed to inhaling secondhand smoke at home when they were children. Individuals who were living with more than one smoker throughout childhood and adolescence were more than twice as likely to develop lung cancer later in life than individuals from smoke-free homes. Other forms of cancer are also significantly increased as a result of living with a smoker.

Smoking may contribute to heart disease. Heart researcher Stanton Glantz concludes from his research that passive smoke causes ten times as much heart as lung disease, making it the nation's third largest killer after active smoking and alcohol abuse. Cancer-causing chemicals in tobacco smoke encourage the development of blood clots in the arteries that nourish the heart.

What Can a Parent Do?

Give up smoking and prevent others from smoking anywhere in your home. According to the 1986 Surgeon General's report, sending smokers to a different room is not enough—it may reduce but doesn't eliminate exposure to tobacco smoke.

FORMALDEHYDE

Formaldehyde is a colorless gas with a pungent odor that is recognized as a probable human carcinogen. It is released from various sources in the home, primarily pressed-wood products and urea-formaldehyde (UF) foam insulation, to a lesser extent carpets, combustion appliances, tobacco smoke, and clothing. The passive release of formaldehyde to indoor air is called offgassing or outgassing.

The Effects on Children

Formaldehyde causes health effects ranging from minor respiratory irritation and watery eyes to possibly cancer. As much as 10 to 30 percent of the population may be sensitive, but sensitivity varies widely from one individual to another. Low concentrations of formaldehyde in the air can cause burning eyes, nose and throat irritation, and headaches, although some people exhibit no symptoms. Higher concentrations can cause severe symptoms like coughing, constriction of the chest, wheezing, and dizziness, and may trigger asthma attacks in asthmatics. The onset of these symptoms is usually associated with moving, remodeling, and purchasing new furniture. Symptoms tend to diminish or disappear when people leave the home. Another symptom, skin irritation, can be caused by contact with formaldehyde-containing clothes, cosmetics, and laundry products.

Formaldehyde is known to cause nasal cancer in rats and mice. For this reason it is considered a possible human carcinogen. Some studies of workers exposed to formaldehyde on the job show a significant association with cancer of the respiratory tract. This raises the possibility that some people may be getting cancer from formaldehyde in the wide variety of sources in the average home.

There is growing concern that repeated exposure to formaldehyde and other volatile organic chemicals in indoor air may cause individuals to become highly sensitive to these substances. The condition, referred to as multiple chemical sensitivity, can be extremely debilitating, forcing sufferers to flee environments associated with modern life (refer to chapter 5, "Maintaining a Chemical-Free Yard," for further discussion). Some people also suffer transient symptoms collectively known as sick-building syndrome when living in poorly ventilated homes that emit a lot of toxic organic vapors like formaldehyde. Symptoms include eye irritation, sleepiness, nausea, irritability, and forgetfulness, all of which diminish or disappear when the person leaves home. This syndrome was first recognized in people that worked in poorly ventilated office buildings.

Reducing Formaldehyde in Pressed Wood

The most common source of household formaldehyde is the adhesive in pressed-wood products like particleboard, hardwood plywood, and medium-density fiberboard. Plywood, found in wall paneling and furniture, is composed of several thin sheets of wood glued together with UF adhesives. Particleboard and fiberboard, used variously for subflooring, furniture, cabinets, and drawers, are composed of wood chips or shavings impregnated with UF resins and pressed into final form. Medium-density fiberboard contains the largest amount of UF, plywood the least.

Formaldehyde outgasses rapidly after manufacture and then more slowly over a period of months or years as the glue breaks down through contact with heat and moisture. Conditions like high humidity and temperature increase the rate of outgassing.

Over the last decade, new manufacturing practices have reduced the amount of outgassing from pressed-wood products. Concentrations are still significantly high in some mobile homes, which are built tight and loaded with pressed-wood products. High concentrations can also occur in conventional homes containing many formaldehyde products, especially where ventilation is poor and humidity and temperature are high.

Reducing Formaldehyde in Insulation

In the 1970s, urea-formaldehyde foam insulation (UFFI) was used extensively as thermal insulation in the walls of already existing homes. A compound composed of urea and formaldehyde was mixed with a foaming agent to give it the consistency of shaving cream. This material could be conveniently injected into wall cavities through small holes that were later sealed. After hardening, the substance was an excellent insulator. It wasn't until after thousands of consumer health complaints came rolling in to the Consumer Product Safety Commission (CPSC) that, in February 1982, it voted to ban UFFI. In 1983, the Formaldehyde Institute had the ban overturned and UFFI became legal again. By that time, however, negative publicity had all but dried up the market for the product.

Half a million homes in the United States carry the legacy of UFFI in their walls. How do you know if your home is one? If your home was insulated prior to 1973, it probably does not contain UFFI. If it was insulated in the 1970s, check with your builder, your insulation company, or any previous owner who occupied it during that period. One of them should be able to help you determine if your walls were insulated with UFFI. If insulation was blown into your walls, don't automatically assume it was UFFI, as cellulose insulation is also applied in this manner. You may see telltale signs of UFFI, for example, a crusty or powdery substance below electrical outlets or pipes entering walls.

If the insulation job was properly done, your walls are sealed and outgassing is low. Take heart—because outgassing declines gradually, formaldehyde contamination of your indoor air may be minimal. Still, in some homes outgassing may only now be occurring; some UFFI foams start to deteriorate after exposure to high humidity and temperature.

Other Sources of Formaldehyde

Other products found in the typical American home can release formaldehyde to indoor air. These sources are insignificant taken alone; in combination with pressed-wood products or UFFI, they can contribute to a dangerously high formaldehyde level. Formaldehyde is used in fabrics as a binder for pigments, as a fire retardant, and to impart stiffness, wrinkle resistance, and water repellency. It can be found in cotton and cotton blends that are used in a variety of products from clothing to draperies and upholstery. Formaldehyde also imparts increased water resistance to products like grocery bags, paper cups, paper plates, waxed paper, facial tissue, napkins, paper towels, and sanitary napkins. It is found in an array of consumer products from cosmetics to fabric softeners, and is released during combustion from gas stoves, gas or wood space heaters, and cigarettes.

What Level of Formaldehyde Is Dangerous?

This may vary from individual to individual especially with regard to respiratory effects. Outdoor concentrations range from 0.002 to 0.006 ppm in remote locations to 0.01 to 0.05 ppm in industrial areas. Older conventional homes have less than 0.05 ppm on average; mobile homes and homes containing lots of pressed wood or UFFI have levels from 0.02 to as high as 4 ppm. About 30 percent of the population would experience the characteristic symptoms if exposed to an amount ranging from 0.5 to 1.5 ppm. Infants, the elderly, and those with allergic and respiratory problems are most at risk. The U.S. government does not regulate formaldehyde levels in homes. However, several states, some European countries, and Canada recommend a level of no higher than 0.1 ppm. If your indoor formaldehyde concentration is greater than this, you may want to take action, especially if family members are experiencing acute symptoms.

Tips for Felling Formaldehyde

The following suggestions should help you control a formaldehyde problem:

• Test your air for formaldehyde. Several companies that sell detectors for formaldehyde as well as other harmful gases are mentioned in the Resources. After exposure to air, most such detectors are sent to a laboratory for analysis.

• Seal UFFI-containing walls carefully, including all holes and cracks, with caulk or spackling compound. Electrical outlets should also be sealed. A special gasket can be purchased at your local hardware store and installed under the cover plate to slow the flow of formaldehyde into your living space. Contact an electrician for advice on this product. The seam between the floor and wall must be tightly sealed with acoustical sealant, caulk, foam-backed tape, or weatherstripping.

• Ventilate your home as often as the weather permits. Maintain moderate indoor temperatures and humidity with an air conditioner and dehumidifier as well as exhaust fans in bathrooms and laundry rooms.

• Cover the surfaces of UFFI-containing walls or pressed-wood products to reduce outgassing. Mylar or vinyl wall coverings or good-quality paint can be tried. Applied in several coats, polyurethane or lacquer may be effective in blocking formaldehyde release from wood products. Cover all surfaces, including the edges. Vinyl carpet or vinyl floor tiling can be used on pressed-wood flooring. The efficacy of these various remedies has not been extensively documented.

• Reduce formaldehyde emissions with ammonia fumigation. Ammonia removes formaldehyde from air by combining with it. Ammonia hydroxide is placed in shallow pans in each room, and the home is sealed for a minimum of twelve hours with the thermostat set at 80° F. During the procedure, fans should be circulating indoor air. In one study of mobile homes, ammonia fumigation resulted in a 45 to 90 percent reduction of formaldehyde. Since working with ammonia can be dangerous, experts should be consulted. Search the "Laboratories" section of your yellow pages. You may need to do some calling around.

• When building a new home, specify that wood products made for the outside of the home—softwood plywood, waferboard, exterior particle-board, and so on—be used indoors. These products contain a waterproof phenol formaldehyde adhesive that outgasses very little compared with nonwaterproof UF adhesives.

• Before purchasing pressed-wood products, ask the manufacturer about the formaldehyde content and what type of adhesive was used. You may want to avoid products containing pressed wood.

• If a child develops skin irritation after exposure to permanent-press material, substitute material like pure cotton or wool (not wool treated for moths) for clothing, and flannel for sheets and pillow cases. Avoid the popular "percale" sheets, which are treated with formaldehyde to make them permanent press (see Resources for untreated textiles).

• To remove formaldehyde from indoor air, use a purifier that contains chemically active materials like potassium permanganate, which bind the gas. Normal air cleaners are useless against formaldehyde and other chemical vapors. For further information, order the booklet "Residential Air-Cleaning Devices: A Summary of Available Information" from the EPA (see Resources).

You may want to remove formaldehyde from your home if your child is still experiencing acute symptoms after the strategies above have been tried. UFFI can be removed from walls, but this costs thousands of dollars and requires an expert to take the walls down. Removing pressed-wood products like cabinets, furniture, and paneling is easier. Of course, you and your family may consider moving to a new home, especially if a family member exhibits a strong sensitivity to formaldehyde. If no family member is sensitive but your indoor concentration is higher than 0.1 ppm, you may still want to take action; formaldehyde, after all, is a suspected human carcinogen.

For more information about formaldehyde, contact the EPA Toxic Substance Control Act assistance line.

CHEMICALS FROM CARPETING

Concern over carpeting was generated in 1987–88 when new carpet installed at EPA headquarters in Washington, D.C., was implicated in illness suffered by employees. Symptoms included eye and respiratory irritation, headaches, dizziness, lack of concentration, nervousness, blurred vision, and nausea. Some people developed multiple chemical sensitivity. In January 1990, the National Federation of Federal Employees filed a petition on behalf of the EPA employees implicating the chemical 4-phenylcyclohexene (4-PC). As a result, the EPA has requested that carpet manufacturers voluntarily test their carpeting for emissions of all kinds.

The chemical 4-PC is emitted from the latex backing used on most carpeting for the last forty years. It is responsible for the distinctive smell of new carpeting. Levels do fall over time; several months after installation the amount of 4-PC outgassing is minimal. 4-PC's effect on health has not been adequately studied.

There are a host of other carpet chemicals that can lead to health problems. Most new carpeting is made from synthetic petroleum-derived fibers, then treated to make it stain- and fire-resistant. Pesticides may be added as well. The foam pad under carpeting can outgas, as can adhesives used for gluing carpets down. Natural carpeting made of wool or cotton may outgas because of dyes and mothproofing chemicals. Older carpeting also has its drawbacks. As carpet ages, the fibers disintegrate and pollute the air with respirable particles.

Here are a few tips for minimizing the hazards from carpeting:

• If your small children have allergies or asthma, don't let them play on carpeting.

• Instead of carpeting, choose a flooring like tile, wood, linoleum, or terrazzo.

• Apply Carpet Guard by AFM (see Resources) to inhibit outgassing. This is an odorless chemical that makes carpets water-repellent, inhibits bacterial growth, and blocks the release of pungent odors from new carpets.

• If you do want carpeting, choose natural fibers—cotton, wool, or a blend of these—without jute or latex backing (see Resources). Don't mothproof wool carpets. If you choose synthetic carpeting, at least ask if it can be unrolled and aired out for several days in the factory before installation. Increase ventilation to rooms for as long as you can smell the carpet.

WAYS TO VENTILATE

In the tightly built, well insulated modern home, outside air may not infiltrate fast enough to dilute air contaminants like CO and formaldehyde. Venting appliances outdoors and using exhaust fans are excellent ways to get rid of pollutants before they spread through your home.

Take note of the fact that exhaust fans can depressurize your home (lower the indoor air pressure). If there is no source of replacement air (an open window, for instance), air may be sucked in from the soil under your home. "Soil gas" is sometimes contaminated with radon (see chapter 2) or pesticides that have been applied to the soil around your home. Air can also be pulled in from open holes like chimneys (backdrafting), bringing dangerous combustion gases into your living space.

To avoid depressurization, open a window in the room where an exhaust fan is in use, or introduce outdoor air to the cold-air return duct of your forced-air heating system. In homes without a forced-air system, individual vents or ducts can supply outdoor air to each room. A more elegant solution, which combines venting stale air, bringing in fresh air, and heating it all in one unit, is heat-recovery ventilation (HRV). HRV units use warm or cold exhaust air (from inside your home) to heat or cool incoming outdoor air. They are most cost effective when used in cold climates where opening windows is impractical.

RESOURCES

Products and Services

AFM Enterprises, 1140 Stacy Ct., Riverside, CA 92507. Telephone: (714) 781-6860.

> Sells Carpet Guard, a water-resistant film as well as odor barrier to reduce vapor emissions from carpeting.

Air Quality Research International, P.O. Box 14063, Research Triangle Park, NC 27709. Telephone: (800) 242-7472.

> Markets home test kit for NO_2 and formaldehyde monitors.

BDC Electronics, P.O. Box 4996, Midland, TX 79704. Telephone: (800) 221-8564.

> Markets the Ultralert System, an alarm that warns of dangerous levels of CO.

Ecology By Design, 1341 Ocean Ave., Suite 73, Santa Monica, CA 90401. Telephone: (213) 394-4146.

> Markets 4-PC-free nylon, wool, and cotton carpeting.

Pace Chem Industries, 779 S. La Grange Ave., Newbury Park, CA 91320. Telephone: (805) 499-2911.

> Sells acrylic sealers for wood furniture, cabinets, and floors to reduce outgassing of formaldehyde.

Pro-Tek Systems, 95 Brownstone Ave., Portland, CT 06482. Telephone: (203) 342-2306.

> Offers NO_2, SO_2, and formaldehyde detectors.

Pure Podunk, RR1, Box 69, Thetford Center, VT 05075. Telephone: (802) 333-4505.

> Sells formaldehyde-free cotton and wool bedding for children. Call for catalogue.

Seventh Generation, Colchester, VT 05446. Telephone: (800) 456-1177.

> Call for their catalogue, which offers cotton bedding and clothing made without formaldehyde.

Testfabrics, 200 Blackford Ave., Middlesex, NJ 08846. Telephone: (908) 469-6446.

> Sells, and provides information about, formaldehyde untreated fabrics.

3M, Occupational Health and Environmental Safety Division, 3M Center Building 220-3E-04, St. Paul, MN 55144. Telephone: (800) 328-1667.

> Markets a passive formaldehyde monitor, sold through local distributors.

Sources of Information

Chemical Exposures: Low Levels and High Stakes. Nicholas Ashford and Claudia Miller. New York: Van Nostrand Reinhold, 1990.

A comprehensive resource on chemical sensitivity illness. It is available from NCEHS (see below) for $16.00.

Consumer Product Safety Commission, Washington, D.C. 20207. Telephone: (301) 492-6580.

Can send you an informative booklet called "An Update on Formaldehyde."

EPA Public Information Center, 401 M St. SW, Washington, D.C. 20460.

Has the following free booklets: "Buying an EPA-Certified Woodstove"; "Residential Wood Heaters Certified by the USEPA" (list of certified wood stoves); "The Inside Story: A Guide to Indoor Air Quality"; "Residential Air-Cleaning Devices: A Summary of Available Information;" and Indoor Air Facts: Sick Building Syndrome, No. 4; Environmental Tobacco Smoke, No. 5; Report to Congress on Indoor Air Quality, No. 6; Residential Air Cleaners, No. 7.

Human Ecology Action League, P.O. Box 49126, Atlanta, GA 30359. Telephone: (404) 248-1898.

A nonprofit organization providing information on health effects of chemicals.

Indoor Air Quality and Human Health. Isaac Turiel. Stanford: Stanford University Press, 1985.

Provides general information on indoor air pollution and its health effects.

"Indoor Air Quality Notes: Residential Formaldehyde Control."

For a copy, write Thad Godish, Department of Natural Resources, Ball State University, Muncie, IN 47306.

National Center for Environmental Health Strategies (NCEHS), 1100 Rural Ave., Voorhees, NJ 08043. Telephone: (609) 429-5358.

An educational, research, and advocacy organization that focuses on chemical sensitivity disorders. Members receive a quarterly newsletter, *The Delicate Balance*, which tracks the latest indoor air issues.

Office on Smoking and Health, Centers for Disease Control, 5600 Fishers Ln., Park Building, Rm. 1-16, Rockville, MD 20857. Telephone: (301) 443-5287.

A clearinghouse of information on the health effects of smoking. Write for a publications list.

Your Home, Your Health, and Well-Being. David Rousseau et al. Berkeley: Ten Speed Press, 1988.

An informative book that reviews many sources of home pollution.

CHAPTER 14

Battling Bio-Nasties

The Problem. Household air is filled with biological particles like pollen, house dust mites, and mold spores.
The Risk to Kids. Excessive exposure to biological pollutants at an early age may lead to the development of allergic diseases like hay fever and asthma.
What to Do. Keep humidity levels low, repair leaks, and dry out wet furnishings. Also, control dust levels.

Seeing your child go through an asthma attack can be a frightening experience. This happened to the mother of ten-year-old Ashley Williams. Before the age of three, when she was diagnosed with asthma, Ashley had endured regular episodes of coughing, wheezing, and shortness of breath. But this time was different; she was desperately gasping for breath. Ashley's mother rushed her to the emergency room and nervously waited until her daughter finally began to respond to medication. It would take three more days before her breathing went back to normal.

Asthma attacks like Ashley's can be triggered by a variety of stimuli, including exposure to biological contaminants. The air in even a relatively clean home contains a surprising number of biological contaminants. These "bio-nasties" include bacteria and viruses, molds and mildew, animal dander (skin scales), saliva and urine, house dust mites, pollen, and insect feces and body parts. Most people are oblivious to the invisible world of microbes. It can be a major problem for people with allergies, asthma, or other respiratory diseases.

Many of the improvements in our indoor environment over the last decades have created conditions ideal for bio-nasties. Central heating distributes bio-nasties throughout the house, tighter insulation increases the humidity in which they thrive, wall-to-wall carpeting gives them lodging, and cold-water detergents won't kill them. All of which means that more and more microbes are being breathed in by our unsuspecting children.

If your house has a bio-nasty plague, there may be several signs: smelly or stuffy air; moisture condensing on windows; moldy books, carpeting, walls, or furniture; and health symptoms like eye, nose, and throat irritation or asthma that are strongest when family members are at home.

THE EFFECT ON CHILDREN'S HEALTH

Some bio-nasties cause noncommunicable diseases like hay fever, asthma, and hypersensitivity pneumonitis. These are different from bio-nasties (like bacteria and viruses) that cause communicable diseases such as influenza and the common cold.

Allergic Diseases

Allergic rhinitis (hay fever) and asthma tend to occur in individuals who produce a large amount of an antibody called IGE after exposure to allergenic particles like pollen or mold spores. These people have been exposed to the allergen before and have become sensitized. IGE combines with allergen and sets off a chain reaction in which tissue-damaging proteins like histamine are released. This causes the symptoms of allergic disease: sneezing, watery eyes, stuffy and runny nose (allergic rhinitis), wheezing, coughing, or breathing difficulty (asthma).

Allergic rhinitis usually develops in childhood, persists into adulthood, and declines in old age. It is uncommon in the very young, occurring in about 3 percent of children under four. About 21 percent of college students exhibit symptoms. Asthma, on the other hand, often afflicts the very young—about 84 percent of asthmatics experience the onset of symptoms before their fifth birthday. It is the major cause of pediatric admissions to hospitals and of school absenteeism. Asthma is defined as a partial narrowing of the airways that can be reversed (not permanently) either by treatment or spontaneously. Both asthma and allergic rhinitis show a familial tendency and often coexist in the same individual.

Recent research suggests that in addition to aggravating the disease in asthmatic children, excessive exposure to foreign proteins of young children who don't have the disease may actually cause asthma to develop. Asthma appears to be growing more prevalent in the United States and other parts of the world.

Hypersensitivity Pneumonitis

This disease is characterized by symptoms ranging from recurrent episodes of breathlessness to flulike symptoms of malaise, cough, headache, and fever. The symptoms occur several hours after inhaling dust containing a wide variety of foreign substances. One form of the disease, called ventilation pneumonitis, is caused by inhaling microbes growing in contaminated ventilation systems. A similar disease is caused by microbes that grow in humidifiers. The symptoms of hypersensitivity pneumonitis usually resolve within a day. However, the disease can become chronic with prolonged exposure to offending microbes, in which case permanent lung damage can occur.

SOME COMMON ALLERGENIC BIO-NASTIES AND THEIR SOURCES

Pollen

The tiny powderlike spores of seed plants originate mostly outdoors and enter your home through doors and windows. Pollen arises from trees, weeds, grass, and flowers with the highest concentrations when weather is warm and dry. Ragweed, one of the most common causes of hay fever, is most abundant in early fall.

Mold Spores

These originate either outside or inside. Outdoors, spores are prominent throughout the growing season in temperate climates, with the highest levels in late summer and autumn during hot breezy spells. Many different molds that produce allergenic spores will grow indoors if there is sufficient water and a food source. Water is plentiful in most bathrooms, in many basements where there is flooding or water leakage, on window sills, and in laundry rooms.

Molds are not that particular about their food. Virtually any carbon source, when damp, supports mold—leather, paper, plastic, wood, and cloth. Unless relative humidity is very high, greater than 75 percent, these materials aren't normally wet enough to support growth. However, a small leak in a roof or pipe or excessive moisture in a bathroom can dampen a surface enough to harbor mold. Soap and grease films on bathroom surfaces can host a variety of molds.

House Dust Mites

Mites are the main allergenic culprits in house dust. They are barely visible to the naked eye; about seven thousand of them could fit on your fingernail. They breed in spring and foam-rubber mattresses, pillows, pajamas, carpets, and upholstered furniture, their main source of food being the scales of human skin. Though harmless, their fecal matter, which sticks to dust particles, can elicit an allergic response in sensitive individuals. Children often develop an allergy to mites when transferred from a crib to a bed. (Cribs and baby carriages tend not to harbor the creatures in great numbers.) Mites flourish in damp, warm conditions and are therefore of particular concern in places like the Pacific Northwest. Their numbers decrease when winter descends because they can't tolerate the low humidity caused by central heating.

Animal Dander and Saliva

These allergens, especially from domestic cats and dogs but also from rabbits, hamsters, gerbils, guinea pigs, horses, mice, rats, and parakeets, are found in house dust. Feathers in bedding, clothing, and furniture have allergenic properties that tend to increase with the age of the material. Feces and body fragments from insects—cockroaches, houseflies, bedbugs, carpet beetles, and spiders—have been implicated in human allergies as well.

An estimated 10 to 15 million Americans are allergic to cockroaches, and not just live ones—cockroach shells, feces, blood, and digestive enzymes can all elicit an allergic response in a sensitized individual. Some 60 percent of asthmatics are sensitive to cockroaches. Allergy to cockroaches may predispose an individual to allergy to other arthropods like crabs, lobsters, shrimp, and crayfish. Therefore, preventing the first allergy is critical. Scientists are now trying to identify cockroach and house dust mite proteins; immunization against them may eventually be available.

HOW ARE BIO-NASTIES SPREAD?

Bio-nasties affect health when they become airborne and are inhaled. Activities like dusting, vacuuming, sweeping, scrubbing, and bed-making stir them up. The slightest air current can send fungal spores swirling through the air. Some fungi produce their spores on long stems that project above the surface on which they grow; they are particularly well positioned for dispersal by the slightest movement of air.

Bio-nasties can also be spread around the home by a forced-air heating system (microbes grow within the ductwork), by air currents from radiant-heat convection, and by air conditioners (their constant supply of condensed water is an ideal breeding ground for bacteria and molds).

Many modern appliances have water reservoirs which, when not absolutely clean, breed microbes: humidifiers, vaporizers, self-defrosting refrigerators, clothes dryers, and toilets. These appliances also have a built-in system for dispersing microbes. Self-defrosting refrigerators have an evaporation pan underneath to collect water. In older, less efficient refrigerators, water accumulates in the pan and breeds microbes which are then dispersed as currents of warm air from the refrigerator's heat exchanger flow over the pan. Clothes dryers pose a problem if the exhaust is vented indoors. Microbes that can withstand heat grow on the internal surfaces of the exhaust pipe, using accumulated lint as a food source. Spores from these organisms are then forced into the air when the dryer is on. As for humidifiers, according to a recent government report, the ultrasonic and cool-mist types emit the greatest number of bacteria and fungi. Steam and evaporation humidifiers as well as those attached to furnaces release an insignificant number.

WHAT CAN A PARENT DO?

If you want to protect your children's health and to prevent damage to your home, here are some precautions:

• Keep indoor humidity levels at around 30 to 50 percent. This will prevent water from condensing on structural materials and furnishings, and eliminate microbe breeding grounds. To control humidity, install exhaust fans vented to the outdoors in bathrooms, the kitchen, and the laundry room. Particularly in summer, use dehumidifiers and an air conditioner. Repair all water leaks.

• Since water-damaged carpets and their underpads as well as furnishings and structural materials harbor mold and bacteria, they should be thoroughly dried and cleaned within twenty-four hours of getting wet. Do not lay carpeting until water problems are taken care of.

• Ventilate attic and crawl spaces to prevent moisture buildup. This will help keep the population of cockroaches and molds low. Covering the earth in crawl spaces with plastic sheeting eliminates them as mold sources. Also, keep your basement as dry as possible by repairing cracks in the foundation through which groundwater can seep.

• Maintain your lawn; a buildup of organic debris can increase the level of mold spores inside. A house that is too well shaded may fall victim to mold spore contamination. Composting plant materials near the house is a special menace to mold-sensitive individuals; mold flourishes in composts.

• If you use a cool mist or ultrasonic humidifier in your child's bedroom, clean the water tray and fill it with fresh, distilled, or demineralized water daily. Do not use tap water in these appliances, as the mineral deposits it contains can build up in the water reservoir and provide a good breeding ground for bio-nasties. Humidifiers can be a major source of mold spores and bacteria, which aggravate asthma and allergic rhinitis and may induce hypersensitivity pneumonitis.

• Evaporation trays in air conditioners, dehumidifiers, and refrigerators should be cleaned frequently. Baths, showers, and other damp surfaces can be treated periodically with a bleach solution to discourage the growth of mold.

If your child has allergies, asthma, or some other respiratory problem, the following tips are also helpful:

• Seek the advice of a physician experienced in allergy diagnosis and treatment if you suspect your child has allergies (see Resources).

• Hot water, electric, and other types of radiant heating systems are best for allergic children because they don't require the movement of a large volume of air. If you have a forced-air system, regularly change the mechanical filter on the cold air return, which filters out most mold spores and pollens. House dust allergens (mites and animal dander) are generally too small to be filtered.

• Keep your home as dust-free as possible. Start by limiting carpeting, stuffed furniture, and dust-catching knickknacks. Dusting furniture with a damp cloth and wet-mopping floors limits the amount of airborne dust. Your vacuum should be outfitted with filter-paper bags rather than cloth bags. If possible, use a central vacuum system that vents outdoors. Allergic family members should not be around during vacuuming or bed-making.

• Keep in mind that the child who has a large number of stuffed animals is going to be exposed to a large number of house dust mites.

• Encase bedding like pillows and mattresses in zippered, impermeable plastic covers to keep the mite population low. Blankets and bedspreads should be washed in hot water monthly. A waterbed is preferable to a box spring and mattress. Throw out foam-rubber pillows that have turned yellow from mold growth.

• If a child is allergic to the family pet, consider giving up the animal, or at least keep it out of living rooms and bedrooms. The allergic child should never groom the pet.

• You may want to install an air-cleaning system in your home (see below for details).

• Discourage your allergic child from playing on carpeting, an excellent home for mites and microbes. They thrive on the bits of food and dander that become trapped in the deep fibers. Don't assume vacuuming will rid your carpets of these pests. Instead, it may simply be pulling microbes from the depths to the surface of the carpet, right under the nose of your playing child.

SHOULD YOU USE AN AIR CLEANER?

Air-cleaning technology is still in its infancy. The EPA ranks it third behind source control (eliminating a source) and ventilation in reducing indoor air pollution.

There are three types of air cleaner on the market: mechanical filters similar to the filter on your furnace, electronic appliances like electrostatic precipitators, which trap charged particles using an electrical field, and ion generators that charge dirt particles, causing them to cling to walls, floors, draperies, or a charged collector. All three types can be installed directly in a central heating and air conditioning system or stand alone as a portable unit.

Air cleaners remove smaller airborne pollutants like tobacco smoke. They are not as effective at removing larger particles—pollens, mite feces, dander, and some mold spores—which are too large to stay airborne for long.

For further information on air cleaners, order the EPA's "Residential Air-Cleaning Devices: A Summary of Available Information" (see Resources).

RESOURCES

Products and Services

Fisons, Environmental Dept., P.O. Box 1766, Rochester, NY 14603. Telephone: (800) 999-MITE.

> Markets the Acarex Test Kit for detecting house dust mites and Acarson for eliminating mites from carpet.

Sources of Information

"Air Purifiers." *Consumer Reports* (February 1989): 88–93.

American Academy of Allergy and Immunology, 611 East Wells St., Milwaukee, WI 53202. Telephone: (800) 822-ASMA.

Helps people find local allergy specialists.

Asthma and Allergy Foundation of America, 1717 Massachusetts Ave. NW, Suite 305, Washington, D.C. 20036. Telephone: (202) 265-0265.

Supplies information on allergic diseases.

EPA Public Information Center, 401 M St. SW, Washington, D.C. 20460.

Provides the following helpful booklets: "The Inside Story: A Guide to Indoor Air Quality"; "Residential Air-Cleaning Devices: A Summary of Available Information," and "Indoor Air Facts, No. 7: Residential Air Cleaners," No. 8: "Use and Care of Home Humidifiers."

The Healthy House: How to Buy One, How to Cure a Sick One, How to Build One. John Bower. New York: Lyle Stuart, 1989.

Among other information, discusses house molds and how to minimize them.

Publication Request, Consumer Product Safety Commission, Washington, D.C. 20207. Telephone: (800) 638-CPSC.

Provides "Safety Alert on Humidifiers" and "Biological Pollutants in Your Home."

CHAPTER 15

Avoiding Electromagnetic Fields

The Problem. Although long considered harmless, extremely-low-frequency electromagnetic fields emitted by electric objects are now suspected of promoting cancer.

The Risk to Kids. Modern, wired-up America has placed children in contact both indoors and out with electromagnetic fields from appliances and power sources, possibly increasing their risk of cancer.

What to Do. Until more is known about this problem, parents should protect their children from strong and constant electromagnetic fields. In most cases, this can be accomplished by simple steps like not using electric blankets with high fields and moving a child's bed away from an area where the electric power cable enters the house.

In 1988, sixteen-year-old Melissa was a star player on her high school basketball team. She was a straight-A student who had plans to try modeling and go to college. But the day before her seventeenth birthday, she suffered a seizure on the basketball court. Within a week, she had been diagnosed with an advanced malignant brain tumor. The chances of a young girl like Melissa getting such a brain tumor are about 1 in 50,000. Disturbingly, two malignant brain tumors and a malignant eye tumor developed in people living in homes close to Melissa's, across from a power company substation and its high-current wires.

Americans have doubled their use of electricity several times in the last forty years. Children today have prolonged contact with some of the electronic equipment found in the home—computers, for example. It is a disturbing thought that the electromagnetic fields produced by the flow of electric current could contribute to cancer. Some epidemiologic studies, which look for the causes of human illness, have linked exposure to these fields with cancer, especially leukemia, lymphoma, and brain cancer.

Parents might assume their children are exposed primarily in front of a TV or computer screen, but there are other, less obvious electromagnetic fields with which they may come into regular contact, for instance, electric substations, neighborhood power lines, and secondary wires bringing electricity into the home. Depending on how far back a house sits from power or transmission lines, a child playing about the house and yard may be exposed to a strong and constant field. The child whose bed is close to the spot where a cable brings electricity into the house may be exposed during sleep to a strong field. And the child who sleeps under an electric blanket is exposed to its field.

WHAT ARE EXTREMELY-LOW-FREQUENCY ELECTROMAGNETIC FIELDS?

Extremely-low-frequency electromagnetic fields (referred to hereafter as ELFs) consist of weak electromagnetic radiation generated by a moving electric current. ELFs are much weaker than X-rays, microwave radiation, or even radio waves. The current in an ELF alternates sixty times per second (60 Hertz), and it is possible that anything magnetic, including molecules in living tissue, will oscillate sixty times a second in an ELF. Its alternating current distinguishes the 60 Hertz ELF from some other fields. For example, the magnetic field that envelops the earth, a condition under which life evolved, is steady or direct current—it doesn't oscillate.

An ELF has two component fields, the electric and the magnetic, which have some different properties. For example, while an appliance is plugged in, not on, the electric field is present but not the magnetic field. Also, electric fields are substantially blocked by objects such as trees and walls, while magnetic fields pass through objects and are difficult to shield. Some researchers feel that it is the magnetic field rather than the electric that manages to reach into a home and affect humans. But there is evidence that both fields can alter cell chemistry.

An ELF induces an electric current in a conducting object. This event is usually invisible, but not on one notable occasion. One night a man named John Filipowski, the owner of an upstate New York farm, carried two ordinary 4-foot fluorescent light tubes to the 345-kilovolt transmission lines that crossed his pastureland. As he neared the lines, the tubes began to glow. When he stood underneath the lines, the tubes shone brightly enough for him to read a newspaper. This raises a question: How much current was the ELF from the transmission lines inducing in Mr. Filipowski himself? Human bodies are also conductors. They have strong electric currents of their own, such as those governing heartbeat and existing across cell membranes. ELFs induce small electric fields inside the body, which in turn creates an electrical current in and around cells. This leads to changes in cell chemistry that may affect health.

THE RESEARCH ON HEALTH EFFECTS

Epidemiologic studies and the occurrence of suspicious cancer clusters suggest that ELFs may increase a child's risk of cancer. But the limitations of these studies have prevented them from actually proving that ELFs cause cancer. More epidemiologic studies are under way.

In the 1970s, the epidemiologist Nancy Wertheimer began an inspection of the home environment of children with leukemia. This work, supported by a later study of David Savitz's, showed a link between high-current power lines and an increased risk of cancer, particularly childhood leukemia. A home was identified as high current if it was situated within 130 feet of a large primary (distribution) line or within 50 feet of the first two spans of wire from a transformer (a box sitting on top of a pole).

This information is easier to understand if you are familiar with how electricity is distributed in America. Electricity leaves large and often distant generating plants at levels of 20 kilovolts (kV)—equivalent to 20,000 volts—and travels to a nearby step-up transformer substation, which boosts it to voltages as high as 765 kV. These high voltages are sent along transmission lines that cross very long distances. Before the power can be distributed to homes, it must travel through local step-down transformer substations, where power is reduced to 35 kV or less. Energy remaining as high as 35 kV travels along distribution lines to another step-down transformer, contained in a box that sits atop a street power line pole. Now voltage is reduced to the 115 V supplied to homes by an electric cable.

Suspicious cancer clusters near sources of unusually high ELFs also have raised the question of a link between ELFs and cancer. Paul Brodeur, an investigative journalist and staff writer at *The New Yorker*, recently described some cancer clusters linking families and schoolchildren with sources of strong ELFs. One cluster occurred in the neighborhood of the girl named Melissa mentioned earlier.

It is hard for epidemiologists to prove that emissions from a substation and its lines caused the cluster of cancer to which she belonged; they must first rule out the effect of other carcinogenic factors, for example, the polychlorinated biphenyls (PCBs) associated with power lines and substations. PCBs were used until 1979 as insulating agents in transformer containers. Unfortunately, weathering allowed PCBs to leak out of containers into the environment. Other problems with epidemiologic studies are contradictory findings, statistically marginal results, and failure to establish a consistent dose-response relationship between ELFs and cancer rates. These difficulties have prompted laboratory scientists to look for a biological mechanism by which ELFs could cause cancer.

Experiments have found that while ELFs do not appear to cause mutations in DNA, they do affect some biological functions in laboratory animals. For example, ELFs cause changes in hormone levels, in the binding of ions to cell membranes, and in biochemical processes like protein synthesis. Scientists are looking at how these biological changes could translate into adverse impacts on health, including the promotion of cancer. For example, ELFs lower levels of melatonin in rats, and low levels of melatonin are associated with a greater incidence of chemically induced breast tumors in female rats.

So far, both the epidemiologic studies and the laboratory research efforts have suggested but not proven that ELFs can cause cancer. At present, the EPA refers to ELFs as a possible carcinogen and continues its research.

COMMON ELF SOURCES

Outdoors, ELFs are released by transmission and distribution power lines. These include high-tension transmission lines carried by tall metal towers, high-current neighborhood distribution lines, and secondary wires carrying electricity into homes. ELFs are also released by ground transformers for underground power lines. Indoors, the electric lines entering a house may be grounded to a water pipe, creating ELFs near plumbing. But beware: electrical experts caution homeowners not to disconnect home grounding systems due to the risks of electrocution and fire. Instead, contact a qualified licensed electrician for advice on alternative grounding. Other indoor sources are electrical appliances and electrical equipment: electric blankets, microwaves, fluorescent lights, electric shavers, television receivers, video display terminals (VDTs), and stereo headphones and speakers.

That's a lot of ELF sources. Need you be alarmed by all of them? Probably not. Some ELFs diminish substantially a short distance from the source. Others are strong, but daily exposure is brief. Appliances do not always emit equally strong electric and magnetic fields. Electric blankets, for example, emit a strong electric field, but not as strong a magnetic field, an electric shaver just the opposite.

WHAT TO DO ABOUT ELFS

Government regulatory agencies have not set safety standards for ELFs because there is no definitive information on hazardous doses. Background ELFs, which are considered harmless, average less than o.1 volt (V) and less than 1 milligauss, or mG (a gauss is a unit of measurement for magnetic fields). Studies have linked magnetic fields as low as 2 mG with an increased risk of cancer but have failed to find a direct relationship between increased field strength and cancer risk. So far there has been an effort to avoid increasing public exposure.

To cite an example: in recent years citizens of Wilmette, Illinois became concerned about a plan by the Chicago Transit Authority to renovate an electric train yard near a park frequented by babies and children. The village trustees worked out an agreement with the transit authority that a proposed new substation cannot be used if electromagnetic radiation coming from the renovated yard exceeds current levels. Montana, Minnesota, New Jersey, New York, North Dakota, Oregon, and Florida have placed limits on electric fields near high-tension transmission lines. Florida and New York have placed or are placing limits on magnetic fields. The limits mostly keep emission levels in line with previous levels. (Average electric field strengths at the edge of high-voltage transmission line right-of-ways are 500 V and up, and average magnetic field strengths are 50 mG. Levels rise inside right-of-ways.)

Should You Practice Prudent Avoidance?

A policy called prudent avoidance is advocated by the authors of a booklet called "Electric and Magnetic Fields from 60 Hertz Electric Power: What Do We Know About Possible Health Risks?" (see Resources). Prudent avoidance entails simple, convenient, and inexpensive steps like not using an electric blanket and arranging beds so they aren't near a secondary wire bringing electricity into the house. While you would avoid purchasing a new home near high-tension transmission lines, or high-current distribution lines, you would not move away from a home near existing lines. In the spectrum of possible carcinogens, ELFs are not the most worrisome. Smoking, alcohol consumption, a poor diet, and too much sun are all more dangerous.

In some cases, people have decided to err on the side of caution to reduce exposure to ELFs, even if it meant taking difficult and costly steps. Citizens of Texas, New York, California, and Louisiana have filed lawsuits to force utility companies to delay, reroute, or abandon the construction of power lines. Parents of schoolchildren have also become concerned about the emissions from high voltage transmission lines located near schools. Many schools have been built on land near power lines and substations because power companies often sold land in their right-of-way at inexpensive prices.

How To Reduce Outdoor Exposure

Louis Slesin, editor of *Microwave News,* suggests that homes, schools, and playgrounds should not be built near high-tension transmission wires. *Fortune Magazine* reports that homes containing children, pregnant women, or women who want to get pregnant should be at least 400 feet from a long-distance high tension line and more than 150 feet from a local high-current distribution line. If your home is near these lines or a substation, call your power company and request that they test for ELFs in your house and yard—anywhere your children spend a lot of time. Independent engineering firms and environmental consulting firms will also take measurements for a fee (see Resources). Or you can rent or purchase a gaussmeter and take measurements of magnetic fields yourself (see Resources).

There are no specific criteria for reducing levels, but you might ask the power company about ways it can shield electric fields and ways to help reduce magnetic fields, perhaps by having them cancel each other out. A common procedure for decreasing magnetic field emissions is to pull high-current distribution wires close together with spreaders. Power companies can also spin or wrap together secondary wires that bring electricity into a home, which discourages the current from jumping to plumbing metal and creating a strong magnetic field. Another method is to use nonconductive plumbing pipes made of polyvinylchloride. Power utilities have various other options for reducing outdoor fields, including relocating and burying power lines, and they are researching more methods.

How to Reduce Indoor Exposure to ELFs

Children should be kept from close and prolonged contact with anything emitting a strong field, for instance, electric blankets and dishwashers. And they can be taught some simple "sit back" precautions for using equipment like VDTs that produce moderate fields.

Electric Blankets

Children should not sleep under electric blankets or in electrically heated waterbeds. Electric blankets can be used to prewarm a bed, but the blanket will have to be unplugged, not just turned off, to stop the electric field. Or simply replace the electric blanket with comforters. This advice also applies to pregnant women, since some studies have linked electric blanket use with increased rates of miscarriage, longer-than-normal gestation periods, and a greater chance of brain tumors in their children. Northern Electric, producer of Sunbeam electric blankets, is introducing blankets with reduced ELFs.

Computer and TV Screens

Children should sit a minimum of 3 feet away from television screens, without exception, not even for video games. Some sources suggest a distance of 10 feet. Children should also sit at least 30 inches (about an adult arm's length) from computer monitors. Magnetic fields can be as high as 23 mG 4 inches from the front of a monitor, but only 1 mG 28 inches away. Place the keyboard farther from the computer to get some distance from the screen. Avoid eye strain by enlarging type size. Monochromatic screens generally give off less radiation than color monitors. Some computers emit even stronger ELFs from their sides, backs, and tops. For this reason, children should not sit within 4 feet of a computer. Remember: Office-type partitions and walls will *not* block magnetic fields.

A few computer manufacturers, including Sigma Designs and Mega-Graphics, are working on monitors that will have fewer ELF emissions. Antiglare filters and antiradiation screens block 95 percent of electric fields but not magnetic fields. The only screens that emit virtually no magnetic fields are those with liquid-crystal displays (LCDs), plasma displays, or displays with light-emitting diodes. One LCD on the market is from Safe Computing. Potential purchasers have to weigh cost and screen-display quality. An information packet and other data about VDTs is available from the Labor Occupational Health Program at the University of California at Berkeley (see Resources).

Baubiologie Hardware is now offering an "ELF ARMOR™" unit for use with Macintosh Plus, SE, II, and Classic computers (see Resources). This unit is installed inside your computer around the cathode end of the CRT, and it partially blocks the 60 Hertz field, reducing its range substantially. According to the Baubiologie catalogue, the device reduces the magnetic field to under 1 mG at a distance of 1 foot from the top, sides, back, and bottom of the computer.

ELFs Near Beds

Since children spend a lot of time sleeping, sources of ELFs near their beds should be relocated. Small electric motors produce strong magnetic fields.

Place items like electric clocks, radios, or fans at least 30 inches from the bed. Or use battery-powered versions. Batteries run on direct current, which doesn't produce 60 Hertz fields.

Other Appliances

You probably don't have to worry about fields from electric appliances in the home that your children don't spend a lot of time using—vacuum cleaners, blenders, electric shavers, and so on. But children should not stand close to an operating dishwasher or microwave. Any appliance operated close to the body like a hair dryer should be used in moderation.

A FINAL CAUTION: VLFs AND MICROWAVE RADIATION

This chapter has focused on the possible health effects of ELFs. There are also questions about the safety of higher-frequency nonionizing radiation. These include very-low-frequency electromagnetic radiation (VLF) and microwave radiation. One household culprit is the cellular telephone, which emits VLF. Electrical products can emit more than one kind of nonionizing electromagnetic radiation. For example, VDTs and televisions emit both ELFs and VLFs. Suspicions about VLFs from computer monitors surfaced recently when two young copy editors at the *New York Times* developed incipient cataracts after about a year's work in front of the monitors. IBM sells a monitor (model 8515, color, $950) with reduced VLF emissions tailored to meet Swedish safety standards. Digital Equipment and Sigma Designs also sell monitors that meet the Swedish standard.

Microwave radiation is generated by many sources, from police radar to long-distance telephone equipment; a primary source in the home is the microwave oven. *Good Housekeeping* magazine recently reported that 65 percent of children aged four to twelve use microwaves at home. Microwave radiation in ovens is absorbed by molecules in food, which cause it to warm up. Because this radiation can burn human flesh ovens are tightly sealed, but small amounts can still leak. To minimize leakage, seals should be carefully maintained. Children shouldn't stand next to operating ovens, which may leak microwaves and which also emit strong ELFs.

Children are exposed to many other sources of nonionizing radiation, including radio waves and radar from broadcasting transmitters and towers, CB radios, and radar dishes. There is concern about this and also about the danger to fetuses of ultrasound devices and the health effects of fluorescent lights.

RESOURCES
Products and Services
Baubiologie Hardware, 200 Palo Colorado Canyon Road, Carmel, CA 93923. Telephone: (800) 441-8971 or (408) 625-4007.

Sells the VHS video "Current Switch: How to Reduce or Eliminate Electromagnetic Pollution in the Home and Office," for $49.95. It gives step-by-step information on dealing with exposure to electromagnetic fields. Also sells the TriField gaussmeter, which measures three exposure fields simultaneously, and can detect ELF, microwave, and radio-frequency fields. Cost is $99. Baubiologie also offers the new ELF ARMOR™ unit to reduce magnetic fields in Macintosh computers. Installation should be performed by individuals experienced in internal servicing and maintenance of the Macintosh. Call for more information or the name of the certified installer nearest you.

Environmental Management and Field Testing, 628-B Library Place, Evanston, IL 60201. Telephone: (708) 475-3696.

Takes electromagnetic field readings for homeowners and companies and offers advice on remediation.

Integrity Electronics and Research, 558 Breckenridge St., Buffalo, NY 14222. Telephone: (716) 885-0011; Fax (716) 885-9212.

Rents (for $60 a week) or sells (for $295) the Consumer Model-27 60 Hz Magnetic Field Meter.

Safe Technologies, 145 Rosemary Street, Suite F, Needham, MA 02194. Telephone: (800) 222-3003 or (617) 444-7778.

Rents (for $59.95 a week) or sells (for $145) 60 Hz and VDT gaussmeters. Also sells computer liquid-crystal displays that emit virtually no EMF. Prices begin at $795.

Safe Environments, 2512 9th St., No. 17, Berkeley, CA 94710. Telephone: (415) 549-9693.

Offers regional ELF testing as part of its Home Inspection Service for environmental hazards. Fees run about $100 per hour.

Sources of Information
"Can Power Lines Give You Cancer?" David Kirkpatrick. *Fortune* (December 31, 1990).

A rough guideline for distances between homes and power lines.

Department of Engineering and Public Policy, Carnegie Mellon University, Pittsburgh, PA 15213. Attn: EMF Brochure.

Send $3 for the brochure "Electric and Magnetic Fields from 60 Hertz Electric Power: What do we know about possible health risks?" by Granger Morgan (1989).

The Labor Occupational Health Program, University of California at Berkeley, 2515 Channing Way, Berkeley, CA 94720. Telephone: (415) 642-5507.

Provides an information packet and other data about VDTs.

"The Magnetic Field Menace." *MacWorld* (July 1990): 136–145.
    Examines the threat from computer ELFs.
The Microwave Debate. Nicholas H. Steneck. Cambridge: MIT Press, 1984. Reviews the debate over microwave safety standards. For a copy ($26.25), write to MIT Press, Order Dept., 28 Carleton St., Cambridge, MA 02142.
Microwave News, P.O. Box 1799, Grand Central Station, New York, NY 10163.
    A commercial newsletter with nontechnical reports on the latest scientific and regulatory developments about 60-hertz fields.
"Power Play." *Discover* (December 1989): 62–68.
    Seeing is believing: photo of Mr. Filipowski and his glowing light tubes. Good discussion of early epidemiologic work.

Part 5

The Healthy Cleanup

It is ironic that efforts to clean up and maintain the home and to eradicate indoor pests often create new contamination, spreading harmful chemicals through the air. This is a special threat to children, who spend so much time at home. Many of the chemicals are carcinogens, for example, benzene (found in household solvents), tetrachloroethylene (in dry-cleaned clothing), or p-dichlorobenzene (in moth repellents and toilet bowl cleaners). Recent EPA studies have shown that many of the most hazardous organic chemicals are indoors at up to twenty times the levels found outdoors. No major federal law specifically regulates indoor air pollution. The burden of protecting children falls on parents. When they are disposed of, toxic household cleaners and pesticides contaminate the air, soil, and water. In this way, they come back to threaten your children's health again. The section that follows explores nontoxic alternatives to chemical cleaners, home maintenance products, and indoor pesticides, and explains how best to use, store, and dispose of chemical products you cannot avoid having around.

CHAPTER 16

Choosing Safer Household Products

The Problem. The expansion of the chemical industry since World War II has introduced into homes an unprecedented number of products, some of which contain carcinogens not listed on labels. These are frequently used, and often stored partially open, in homes that may not be adequately ventilated.

The Risk to Kids. Children at home are often near a caregiver who is using chemical cleaning or maintenance products. Therefore, children can be exposed to a high level of toxic vapors, to which their bodies are especially vulnerable. Children's high breathing rate means they will inhale more of the pollutants relative to body size than adults.

What to Do. Parents can reduce their use of hazardous household products by substituting simple homemade cleaners or newly available commercial lines of nontoxic products. Parents can also provide better ventilation when toxic products are in use and store them outside the house once they have been unsealed.

The average American home contains 3 to 10 gallons of hazardous materials, defined by the EPA as flammable, reactive, corrosive, or toxic. Obvious products like paint thinner come first to mind; then there are the less obvious: polish, wood preservative, oil, glue. Many of these materials release chemical vapors, as Andy Bendelow, who runs Ecocleaning Services in Chicago, learned one day when he nearly fainted from the fumes of a conventional tub and tile cleaner. The experience persuaded him to switch to nontoxic products in his business. One EPA-sponsored study found more than 350 organic chemicals in the air of a single home. A typical one is methylene chloride, an animal carcinogen and a suspected human carcinogen. Methylene chloride is used as a solvent in paint thinners and strippers, as a propellent in spray paint and some hair sprays, and in miscellaneous products: wood stains, varnishes, spray shoe polish, artificial snow, water repellent, adhesive, and adhesive remover. The substance makes up a large portion of some products—50 to 80 percent of almost all paint strippers. Few labels list it by name, instead indicating that the product contains chlorinated solvents or aromatic hydrocarbons.

THE PROBLEM CHEMICALS

Hazardous household products fall into four main groups: cleaners and polishes, paints and solvents, auto products, and pesticides. (Pesticides are covered in chapters 5 and 17.) Each group contains dangerous chemicals, from toxic acid compounds to carcinogenic organic chemicals. Charts identifying these chemicals and their health effects (see Resources) as well as other sources were used to compile the information below. Note that corrosive chemicals and vapors wear away the surface of materials and can damage body tissue on contact. Irritant chemicals cause soreness or inflammation of the skin, eyes, mucous membranes, and respiratory system. Toxic chemicals cause injury or death upon ingestion, absorption, or inhalation.

If, after reading about the following chemicals, you are interested in more information, consult the Remote Access Chemical Hazards Electronic Library, also known as RACHEL (see Resources).

Hazardous Chemicals in Cleaners and Polishes

Air fresheners contain formaldehyde, a respiratory irritant and suspected human carcinogen that can also cause nausea, headaches, and shortness of breath. Formaldehyde is used in spray deodorizers and can make up 37 percent of wick deodorizers.

Disinfectants often contain sodium hypochlorite, ammonia, or phenols. Sodium hypochlorite and ammonia can burn skin and irritate lungs. Phenols are flammable and toxic to the respiratory and circulatory system.

Drain cleaners may contain sodium or potassium hydroxide (lye), which burns skin and eyes and may irritate lungs. Other typical ingredients are hydrochloric acid, which also burns skin and has an irritating vapor, and petroleum distillates like trichloroethane. Trichloroethane irritates the nose and eyes and depresses the central nervous system, causing drowsiness or dizziness. It is also a mutagen, that is, it can increase the frequency or extent of mutation.

Toilet bowl cleaners may contain hydrochloric acid, oxalic acid, or chlorinated phenols. Both acids are corrosive and cause burns; chlorinated phenols are very toxic to the respiratory and circulatory system.

Oven cleaners contain sodium or potassium hydroxide.

Bleach contains sodium hypochlorite.

Window cleaners may contain diethylene glycol or ammonia. Diethylene glycol depresses the central nervous system.

Spot removers contain hazardous chemicals including sodium hypochlorite. Others are perchloroethylene or trichloroethane, which cause liver and kidney damage, and ammonium hydroxide, whose vapors are highly irritable to skin, eyes, and respiratory passages.

Floor cleaners and waxes can contain diethylene glycol and petroleum solvents. Petroleum solvents irritate skin, eyes, nose, throat, and lungs.

Furniture polish can contain petroleum distillates or mineral spirits, which irritate skin, eyes, nose, throat, and lungs.

Shoe polishes may contain toxic organic solvents like trichloroethylene and methylene chloride.

Dry-cleaning solvents that remain in clothes may contain tetrachloroethylene, a suspected human carcinogen. Tetrachloroethylene can remain in homes for up to a week after newly dry-cleaned clothing is placed in a closet.

Hazardous Chemicals in Paints and Solvents

Most everyone breathes paint fumes from time to time, but few people know how varied and toxic the evaporating solvents in paint are.

Oil- and enamel-based paints contain toxic pigments and volatile, flammable organic compounds like toluene and benzene. Toluene irritates skin and the respiratory system and depresses the central nervous system. Benzene is a carcinogen.

Latex and water-based paints are considered less hazardous than oil- and enamel-based paints, but they can contain toxic ingredients like resin, glycol ether, ester, pigment, mercury, and fungicides (see chapter 1, "Limiting Lead Paint and Dust").

Rust paints contain methylene chloride, petroleum distillates, and toluene, which are flammable and toxic.

Paint thinners contain the probable human carcinogen methylene chloride and other toxic organic chemicals like acetone. People absorb paint thinners by inhalation and directly through the skin.

Wood preservatives contain some toxic ingredients. One is pentachlorophenol, or penta, found in water-repellent products. Penta, which can cause dermatitis, nausea, headache, and liver damage, is a persistent chemical that has been found in more than 80 percent of the U.S. population. Another preservative, creosote, can burn skin, irritate the respiratory system, and cause headache and vomiting. You may have a deck or porch made of wood that has a greenish tint, so-called wolmanized lumber. This wood is commonly treated with water-soluble salts, and arsenic can leach out of it into soil. This poses a risk to children playing in the soil or crawling on the wood.

Furniture strippers contain toxic and flammable ingredients like toluene and methylene chloride.

Stains and varnishes contain toxic and flammable ingredients including mineral spirits, alcohols, benzene, and gasoline.

A number of cleaning and paint products are sold in aerosol cans. Aerosol cans cannot be recycled and they effectively disperse their toxic contents into the air. They also add a propellant gas to the air, which may be an irritating or toxic chemical like methylene chloride.

Hazardous Chemicals in Auto Products

Although auto products don't usually pollute indoor air, auto exhaust pollutes garage air, and the disposal of dirty motor oil is one of the biggest hazardous waste problems.

Used motor oil turns black and accumulates chemical waste like lead and benzene.

Car batteries contain toxic sulfuric acid and lead. Household batteries contain other toxic metals like mercury and cadmium. They are not recycled and are not generally recognized as hazardous waste.

Antifreeze contains highly toxic ethylene glycol, an organic chemical that can damage skin, blood, and the kidneys. The ingestion of only three ounces can kill an adult.

Transmission fluid contains toxic and flammable hydrocarbons and mineral oils.

Brake fluids contain flammable glycol ethers and toxic heavy metals.

Gasoline fumes from the family car are a major source of exposure to benzene. Children can be exposed to high concentrations at the gas station and also from vapors in the garage. Benzene levels during fill-ups have been measured at 1 ppm, 300 times the normal outdoor level.

WHY ARE CHILDREN MORE VULNERABLE?

While parents are usually careful about preventing their children from ingesting household products, they don't often consider the danger of inhaling them. But children take in more pollutants than adults by breathing and also through skin absorption relative to body size. Children's bodies may be less able to detoxify many pollutants (see chapter 4, "Getting Rid of Toxic Art Materials").

Household Habits that Increase a Child's Exposure

Parents unwittingly increase their children's exposure to toxic vapors when they seal up homes with insulation and weather stripping to conserve energy. Also, parents often neglect ventilation when they use chemical products, even when the label advises it. (According to Dr. Lance Wallace, an expert in the field of indoor air pollution, if you can smell a product, it's not good.) And parents probably underestimate how much air pollution is created by the constant but unnoticed evaporation of chemicals from partly closed containers stored inside.

What Product Labels Don't Reveal

There are two common but incorrect assumptions about chemical household products. The first is that the label lists all ingredients. The second is that the product is safe if it is on the market. Labels that contain ingredients known to carry an immediate or long-term risk are regulated by the Hazardous Substances Act. These labels must include a word like *Warning, Danger,* or *Caution*, but not necessarily a list of the hazardous substances. As mentioned earlier, suspected human carcinogens are not listed by name in products like air deodorizer or aerosol paints. As for the safety of chemicals in marketed products, many have not even been tested for toxicity.

USE SAFER PRODUCTS

Parents can avoid toxic chemical cleaning and maintenance products when alternatives are available. Many relatively nontoxic products can be made at home using a few inexpensive, readily available ingredients. Parents also have the option of purchasing nontoxic products from a number of new lines available at stores or by mail. If you look around your community, you may find a "green" cleaning service like Andy Bendelow's in Chicago.

Make Your Own Nontoxic Products*

Those who have made the switch to natural cleaning products warn that there may be a period of trial and error before others find products to suit them. There are advantages to making your own products: you'll save money and avoid the plastic packaging of commercial lines. The five main ingredients in nontoxic cleaning products can all be purchased at the grocery store.

• Baking soda (sodium bicarbonate) cleans, deodorizes, scours, and softens water.

• Borax cleans, deodorizes, disinfects, and softens water.

• Soap, as opposed to detergent, biodegrades safely and completely. Even phosphate-free laundry detergent contributes to water pollution.

• Washing soda (sodium carbonate) cuts grease, removes stains, disinfects, and softens water. It is available from Arm and Hammer.

• Vinegar (white) cuts grease and freshens. Used full-strength in a squirt bottle, it can clean soap film and mildew off plastic shower curtains. Soaking shower heads in full-strength vinegar removes hard-water buildup, and gently boiling one-half cup each of water and vinegar removes lime deposits on tea kettles.

Other commonly used ingredients are salt and lemon juice. Ammonia can be used when nothing else works, but it burns eyes and skin. Follow label instructions and provide good ventilation.

State EPA charts and the Environmental Hazards Management Institute's "Household Hazardous Waste Wheel" not only identify toxic products but also list recipes for safer products. Recipes are also listed in books by Debra Lynn Dadd and Heloise and in information put out by environmental groups like Greenpeace and Citizens for a Better Environment (see Resources). Here are some examples of nontoxic strategies:

• Air freshener: Set out small bowls of baking soda or vinegar. Sprinkle baking soda in diaper pails. Add fragrance to the air by boiling sweet herbs and spices in water. Don't forget simple strategies like removing the source

---

* Warning: Some cleaning products, if mixed together, will form poisonous gases. *Never* mix ammonia with chlorine bleach, toilet bowl cleaners, rust removers, or oven cleaners. *Never* mix household bleach containing sodium hypochlorite with acidic products like vinegar, with toilet bowl cleaners, or with liquid dishwasher detergent.

of the odor, opening windows, and using an exhaust fan. Grow lots of house plants, which help clean air by absorbing and breaking down toxic organic chemicals like benzene, trichloroethylene, and formaldehyde.

• Disinfectant: Wash items with soap and water, or with one-half cup of borax in a gallon of water.

• Drain cleaner: Use hair traps (drain sieves), add a blockage-loosening solution, rinse with boiling water, and use a plunger or plumber's snake. One loosening solution consists of one cup each of baking soda, salt, and vinegar. Pour these down the drain, let it sit fifteen minutes, and flush thoroughly with boiling water. Another solution is a pound of washing soda in three gallons of boiling water. This is poured down the drain and then a plunger is used. Do weekly maintenance on your drains by pouring one-fourth cup of baking soda down the drain, followed by one-half cup of vinegar. Close the drain until the fizzing stops.

• Toilet bowl cleaner: Use a toilet brush and baking soda. Or pour undiluted white vinegar in and let sit for five minutes. For lime deposits above the water line, apply paper towels saturated with vinegar until the deposits loosen. Be sure to remove the paper towels before flushing.

• Oven cleaner: Use a paste of baking soda and water. First wipe away grease and spills, and scrub charred spots with a bristle brush, then apply the paste. Scrub again after five minutes. Also try scrubbing with steel wool and a mild cleaning powder like Bon Ami. Practice prevention by putting a cookie sheet or foil in the oven to catch spills.

• Laundry bleach: Add white vinegar, baking soda, or borax to the wash.

• Window cleaner: Mix three tablespoons of vinegar with a quart of water for routine jobs. Sometimes plain water is adequate. Prewash very dirty windows with soapy water, remembering to use soap and not liquid detergent.

• Stain remover: Treat stains immediately with a sponge and cold water. There are many recipes tailored to specific stains. Test solutions first on a hidden area to make sure they don't hurt the fabric. Treat coffee and chocolate stains with a solution of a teaspoon of vinegar in a quart of cold water. Sponge ballpoint ink stains with rubbing alcohol, then rub with soap, rinse, and wash. Rub grass stains with glycerine (available in drugstores), let stand one hour, and wash. Treat perspiration with a solution of vinegar or lemon juice in water. Mildewed items should be washed with soap and water and dried in the sun. If a spot remains, treat it with lemon juice, rub it with salt, dry in the sun, and wash. Treat grease stains on cottons by pouring on boiling hot water, then rubbing with washing soda dissolved in water. For grease stains on other materials, dampen and rub with soap and baking soda. Wash in extra hot, soapy water.

• Floor and furniture polish: Mix one part lemon juice with two parts olive or vegetable oil. Alternatives for polishing furniture are beeswax or beeswax and olive oil. Or mix two teaspoons of lemon oil with one part mineral oil in a spray bottle. An alternative for floors and wood cabinets is Murphy's Oil Soap, made entirely of vegetable oils.

• All-purpose cleaner: Mix one-half cup of ammonia (handle with care and provide ventilation), one-fourth cup vinegar, and a handful of baking soda in a bucket of warm water. This can be used on linoleum, walls, and in the bathroom.

• Shoe polish: Use walnut or olive oil. Apply the oil and then buff.

• Dry-cleaned clothing should be aired out for one or two days near an open window, never in the bedroom closet. If your clothes have a strong chemical odor when you pick them up at the dry cleaners, get a new cleaner.

• Nonaerosol products: Use spray bottles instead of aerosol products. For example, mix your own spritz starch using two tablespoons of cornstarch in a pint of cold water. Shake before using.

• Scouring powder: Use soap mixed with baking soda, borax, or salt to replace commercial powders with undesirable ingredients like color, detergent, and chlorine or other bleaches. Or use a milder product like Bon Ami, which contains no bleach or phosphates.

• Laundry powder: Use soap or soap flakes and washing soda, added to the water before the clothes. Be sure to wash your clothes first in plain water to remove detergent residues or clothes may turn yellow.

• Dish soap: Use powdered soap or liquid soap, like Murphy's Oil Soap, plus two or three teaspoons of vinegar for tough jobs.

• Automatic dishwasher soap: This can be made of equal parts borax and washing soda. If your water is hard, try increasing the proportion of washing soda. An alternative is to use a phosphate-free detergent (see Resources).

• Metal cleaner: Rub brass with a mixture of equal parts salt, flour, and a little vinegar. Rub copper with lemon juice and salt, or hot vinegar and salt. Rub silver and stainless steel with a paste of baking soda and water. Always use a soft, damp cloth, rinse, and polish with a dry cloth.

Buying Nontoxic Brands

There are new, safer household products in stores or available by mail order. Alternative maintenance products like paints, varnishes, and glues, difficult to make at home, are sold by a growing number of companies, Sinan and EcoDesign (see Resources) among them.

A number of catalogues now offer natural products. The Seventh Generation Catalogue is free and lists a product line from Belgium called Ecover (see Resources). Among Ecover's products are a hydrogen peroxide fabric bleach and a vegetable-based toilet cleaner. Two other free catalogues offering alternative products are Baubiologie Hardware and NEEDS (see Resources).

One line of natural cleaning products, Earthrite, includes a general-purpose cleaner as well as cleaners for countertops, floors, wood paneling, toilet bowls, and tubs and tiles. These products contain no phosphates, chlorine, petroleum distillates, synthetic dyes, or perfumes. Active ingredients include an acid from fermented corn that breaks up dirt and grease, and soap mixed with an acid distilled from citrus plants. Earthrite products are biodegradable, not tested on animals, and packaged in recyclable plastic bottles. They are available in some supermarkets and hardware stores and by mail order (see Resources). Another line, Life Tree, is available in natural food stores and by mail order (see Resources). It offers dishwashing liquid with aloe vera and calendula, laundry liquid, and Home-Soap, an all-purpose cleaner. These products use vegetable-based oils and avoid phosphates, synthetic colors, and perfumes.

Alternative home maintenance supplies are sold by AFM (see Resources). This company has tailored its products to chemically sensitive individuals. It makes over twenty low-odor products, including Super Clean, a biodegradable all-purpose cleaner, Duro Stain for wood, Vinyl Block to control outgassing, and Water Base Varnish. It also offers tile grout, carpet guard, and spackling compound, plus alternatives to polyurethane and lacquer. They do not reveal product ingredients.

Sinan (see Resources) advertises natural building materials in a catalogue that gives information on applying paint, wood finish, and other products. Sinan products contain no petroleum-based crude oil or plastic ingredients. Products are offered in five main categories: natural impregnation (insect repellent) like pulverized borax and wood pitch, varnish and wax, natural resin lacquer, wall paint and glue, natural cleanser and polish, and plant color for painting and modeling.

Livos Paints is a German line of natural paints (see Resources). Its paints, mostly natural oil–based, contain a synthetic solvent derived from petroleum that the company considers less irritating to the skin than turpentine. Livos sells enamel and flat paint, for both interior and exterior use, plus oil primer, varnish, shellac, stain, and thinner. It also has floor wax, spackler, and adhesive.

RESOURCES

Products and Services

AFM (American Formulating and Manufacturing) Enterprises, 1140 Stacy Court, Riverside, CA 92507. Telephone: (714) 781-6860. Fax: (714) 781-6892.

Sells milder cleaning and maintenance products for the chemically sensitive.

Baubiologie Hardware, 200 Palo Colorado Canyon Rd., Carmel, CA 93923. Telephone: (800) 441-8971 or (408) 625-4007.

Offers many home cleaning and maintenance products, from phosphate-free dish detergent to natural linoleum.

EcoDesign, 1365 Rufina Circle, Santa Fe, NM 87501. Telephone: (505) 438-3448; orders taken at (800) 621-2591.

Offers free "Natural Choice" catalog, and sells Livos Plantchemistry paints and related products made primarily from natural materials.

Environmental Hazards Management Institute (EHMI), P.O. Box 932, 10 Newmarket Rd., Durham, NH 03824. Telephone: (603) 868-1496.

Will send you the Household Hazardous Waste Wheel for $3.75.

Illinois EPA, 2200 Churchill Rd., Springfield, IL 62706. Telephone: (217) 782-6760.

Will send you the Household Hazardous Products Chart (no. 14662-8/85-5000).

Magic American Corporation, 23700 Mercantile Rd., Beachwood, OH 44122. Telephone: (800) 321-6330 or (216) 464-2353.

Offers the Earthrite line of milder, nontoxic home products.

NEEDS (National Ecological and Environmental Delivery System), 527 Charles Ave. 12-A, Syracuse, NY 13209. Telephone: (800) 634-1380.

Has a free mail-order catalogue with various home cleaning and maintenance products.

RACHEL (Remote Access Chemical Hazards Electronic Library), Environmental Research Foundation, P.O. Box 73700, Washington, D.C. 20056-3700. Telephone: (202) 328-1119.

Has a computer database with chemical hazard profiles, offers an information packet to the public.

Seventh Generation Products for a Healthy Planet, Colchester, VT 05446-1672. Telephone: (800) 255-7800.

Offers various home cleaning and maintenance products.

Sinan Company Natural Building Materials, P.O. Box 857, Davis, CA 95617-0857. Telephone: (916) 753-3104.

  Free catalogue featuring Auro paint and paint-related products, made from natural materials only.

Sunrise Lane Products, 780 Greenwich St., New York, NY 10014. Telephone: (212) 242-7014.

  Offers the Life Tree line of milder, nontoxic cleaning products.

Sources of Information

Citizens for a Better Environment, 407 S. Dearborn, Suite 1775, Chicago, IL 60605. Telephone: (312) 939-1530.

  A regional organization with information on alternative household products.

Clean and Green. Annie Berthold-Bond. Woodstock, N.Y.: Ceres Press, 1990.

  485 ways to clean, polish, and disinfect using ingredients like baking soda and borax.

The Earthwise Consumer, P.O. Box 279, Forest Knolls, CA 94933.

  A newsletter with advice on household hazards. It also lists nontoxic household products.

Environmental Health Coalition, 1844 Third Avenue, San Diego, CA 92101. Telephone (619) 235-0281.

  Offers free "Home Safe Home Kit" with fact sheets on house cleaners, auto care, paint, and more. Also offers the "Ecological Buying Guide."

Environmental Health Watch and the Housing Resource Center, 4115 Bridge Ave., Cleveland, OH 44113. Telephone: (216) 961-4646.

  Has information on nontoxic cleaning and maintenance products. Also sponsors the Blueprint for a Healthy House conference series.

Greenpeace, 1436 U St. NW, Washington, D.C. 20009. Telephone: (202) 462-1177.

  Has references for alternative household products.

Heloise Hints for a Healthy Planet. Heloise. New York: Perigee Books, 1990.

  Lists many home cleaning tips and product recipes.

Nontoxic and Natural. Debra Lynn Dadd. Los Angeles: J.P. Tarcher, 1984.

  A "yellow pages" of natural products.

The Nontoxic Home. Debra Lynn Dadd. Los Angeles: J.P. Tarcher, 1986.

  Information on nontoxic cleaning products.

CHAPTER 17

Eradicating Indoor Pests Safely

The Problem. A recent EPA study of thirty-two common household pesticides showed that indoor levels were almost always higher than outdoor levels; some pesticides were present in concentrations ten or more times greater inside than out. The indoor environment is a significant nonoccupational source of exposure to pesticides.

The Risk to Kids. Many common household pesticides are neurotoxic or carcinogenic; a child's developing body is especially vulnerable to them.

What to Do. Parents can eliminate indoor pests with a variety of safer alternatives, ranging from physical barriers and heat to low-toxicity chemicals and natural biological controls.

Steve House worked for over four years spraying homes and buildings for roaches and termites. He developed recurring headaches and dizziness, and his mood swings were so severe that his wife didn't know him anymore. After a bout in a mental health clinic, he was finally properly diagnosed a victim of chlordane poisoning. But Steve was not the only victim. "I felt like Attila the Hun," House lamented. "I had sprayed this stuff around families with children without knowing anything about its toxic properties."

Indoor pesticides, used to control everything from termites and cockroaches to ants and fleas, often end up in the air your children breathe. A 1990 study by the EPA, "Nonoccupational Pesticide Exposure Study" (NOPES), addressed the question of how much pesticide the general public is exposed to at home. Up until this time, little was known about indoor exposure to pesticides, although extensive studies had been done in occupational settings. NOPES showed that the mean concentrations of some of the most common pesticides including chlordane, diazinon, propoxur, and heptachlor, were at least ten times higher indoors than outdoors in the households sampled. This study, an important first step in bringing attention to the danger of indoor pesticides, also points out the need for additional studies on their health effects in such concentrations.

WHY ARE CHILDREN AT SPECIAL RISK?

Children are more vulnerable than adults to indoor pesticides for physiological reasons (see chapter 5, "Maintaining a Chemical-Free Yard" and chapter 9, "Purging Pesticides From Produce") and because they spend so much time at home, close to the floor, where pesticides may concentrate. A study done in 1990 by R. A. Fenske and others, "Potential Exposure and Health Risks of Infants Following Indoor Residential Pesticide Applications," examined infants' exposure to the common household pesticide chlorpyrifos (a neurotoxin), which was sprayed on carpeting for the treatment of fleas. A substantially greater concentration of the chemical was detected near the ground, in the breathing zone of crawling infants, than in the breathing zone of the sitting adult. Window ventilation did not decrease infant-breathing-zone levels as much as it did adult-breathing-zone levels.

There are some epidemiologic studies showing an increased risk of cancer in children exposed to pesticides at home. One done by R. Lowengart and others demonstrated that the risk of childhood leukemia increases as much as fourfold in households where pesticides are used at least once a week.

HOW CAN YOU CONTROL INDOOR PESTS SAFELY?

Hazardous pesticides used for termites, cockroaches, and ants can be replaced with safer products and control strategies that discourage pests from entering the home and breeding. Ask your pest-control company if it carries these products and if it can implement the control strategies discussed below. If you can't find one in your area, contact the manufacturers or distributors (see Resources) of the products mentioned below to find out how they can be obtained and applied. Many of these products are new and may not yet be available for sale everywhere in the United States. Some may only be available by mail order. You may want to provide your local pest-control operator with this information and encourage him to use low-toxicity products. Or you may want to try the products and strategies yourself.

Many of the less toxic products and integrated pest management (IPM) strategies discussed below are recommendations of the Bio-Integral Resource Center, a nonprofit organization that provides literature on controlling indoor pests safely (see Resources).

TERMITES

Horror stories abound regarding the vapors of synthetic termiticides—chlordane, heptachlor, aldrin, and dieldrin—present many years after treatment. These termiticides, known collectively as cyclodienes, were poured or injected into soil around homes, injected through foundations into soil beneath homes, or injected into holes drilled in hollow-block walls. Chlordane, one of the most toxic termiticides, can linger in soil around homes for thirty years. The EPA says that 90 percent of homes properly treated with synthetic termiticides still have detectable levels in the air one year or more after treatment, with basements containing the highest levels. Improper application of termiticides, which includes spraying the soil or wood in crawl spaces (this is illegal in most states) and contaminating indoor surfaces, can lead to much higher indoor levels.

The vapors from these chemicals can cause headaches, dizziness, muscle spasms, tingling sensations, and nausea. Long-term exposure may cause damage to the liver and central nervous system. The EPA banned cyclodienes in 1988 because they are probable human carcinogens and also because they persist in the environment, remaining an inhalation hazard for many years in treated homes. The EPA believes there can be long-term health effects from indoor exposure to termiticides, even from those applied properly. If your home was treated for subterranean termites prior to 1987–88, it is possible that chlordane or one of the other cyclodienes was used. Subterranean termites are responsible for 95 percent of termite damage in American homes.

With the banning of chlordane and heptachlor in 1988, chlorpyrifos (dursban) became the most popular treatment for subterranean termites. Chlorpyrifos is a potent neurotoxin that can linger in indoor air for up to sixteen years. Severe drywood termite infestations, found mostly in southern and coastal regions of the country, are commonly treated by fumigation (tenting the home) with the highly toxic chemicals methyl bromide and vapam (sodium-methyldithiocarbamate). Residents must abandon their home for several days to give the vapors time to dissipate.

Testing for Termiticides

If you can smell chemicals, especially when the furnace or air conditioner is operating, and if family members have some of the symptoms mentioned above, termiticides may have been improperly applied to your home. Vapors could be gaining easy access to living spaces. To determine if termiticides are present, your indoor air can be tested for a cost of $50 to $500. To locate a reputable testing laboratory in your area, contact the National Pesticide Telecommunications Network (see Resources). It can also tell you what the test results mean and where to go for help if need be.

Safer Termite Control

There are several alternatives to synthetic termisticides. Several can be used in combination by you or your pest-control company.

• Sand barriers can be applied around the outer foundation of the home or in crawl spaces to separate termites from structural wood. Subterranean termites can't tunnel through certain grades of sand because the galleries they construct collapse. For information on the type of sand to use and where to get it, contact Isothermics or the Bio-Integral Resource Center (see Resources).

• Hot air can be pumped through flexible ducts into a home until wood materials reach 120° F. This temperature is lethal to most insects, including ants and cockroaches. No damage is done to furniture. For information on heat treatment, contact Isothermics.

• The Electrogun is a hand-held device that emits a strong but harmless electric current. It can be applied as a spot treatment to wood where termite galleries are thought to be. This instrument must be used by a trained pest-control operator.

• Biological controls consist of natural predators such as nematodes (see Resources), which can be applied to the soil around your home. They will enter the termites and kill them. Argentine ants will attack and kill subterranean termites if their mud tubes are broken open for entry.

• Desiccating dusts like diatomaceous earth and amorphous silica gel (see Resources) are low-toxicity materials for controlling termites as well as pests like cockroaches and ants. These dusts are sorptive, meaning they are absorbed by or cling to the outer cuticle of an insect, causing the layer to strip off; the insect dies of dehydration. Dusts are best used in isolated areas like attics for drywood termites. If kept dry, dusts remain lethal for termites for the life of your home. You should avoid breathing these dusts because the particles can irritate the lungs.

COCKROACHES

These pesky creatures come in at least three common varieties: German, brown-banded, and Oriental. The German cockroach prefers warm, moist areas like kitchens and bathrooms and can exist on small crumbs of food. Water to quench the roach's thirst is plentiful around sink traps, leaking pipes, toilet bowls, pet dishes, and flower vases, to name just a few. The brown-banded cockroach can be found throughout the house, even upstairs in bedding and closets. The Oriental roach prefers locations like crawl spaces and basements.

Cockroaches are introduced into a home in the form of eggs deposited on supermarket bags and boxes, furniture, appliances, and other household goods. They can also enter from outdoors, where they live in plant litter and wood debris, through loose-fitting windowscreens, under a door, or through holes in the foundations. Once in your home, roaches take up residence in its dark nooks and crannies. Favorite spots are the back side of cabinets and walls. In such hospitable environments, a German cockroach can live up to five months. During that time a female can give birth to some 240 offspring. Thus in a relatively short time your home can be substantially infested.

The bad reputation roaches have is not entirely justified. There is no evidence that they carry disease. They can transmit bacteria by walking over contaminated surfaces, but so can kittens, dogs, babies, and the bottom of your shoes. In fact, roaches play an important role in nature: they decompose decaying material.

However, for some 15 million Americans that are allergic to cockroaches, getting rid of them is not just an aesthetic exercise. These people can experience upper respiratory problems and skin irritations, have difficulty breathing, go into shock, and in rare cases, die. About 60 percent of asthmatics are sensitive to cockroaches.

Home Modifications

The first step in an IPM strategy for controlling cockroaches is to keep them from entering your home. Repair or replace loose-fitting screens and make sure your outside doors are snug. Examine the outside of your home, especially near the foundation, for small cracks or holes through which roaches, ants, and other pests can gain access. Caulk them or fill them with cement. Also, keep plant debris, firewood, and garbage well away from the home to reduce the population of outdoor roaches. Indoors, repair leaking pipes, and remove sources of free-standing water or food like flower pots and pet dishes. Clean your cat's litter box often. Plug up small holes and cracks between baseboards and walls with caulk. It is also a good idea to caulk around kitchen cabinets and areas where the countertops meet walls. This is no small job! Store food in tight-fitting plastic or metal containers, not in cardboard boxes, and take out the trash often.

Safer Chemical Controls

If you have a persistent problem, you may need to do more than make home modifications. There are a variety of pesticides on the market that are relatively less toxic.

Boric acid, applied as a light dust, has always been one of the most effective pesticides for roaches. Because it is nonvolatile, roaches don't detect it and are therefore not repelled by it. Roaches walk through the dust, pick it up and ingest it, and die seven to fourteen days later. (Unfortunately, many homeowners and exterminators want to see a quick kill.) Roaches have not yet developed resistance to boric acid, despite decades of use. It is generally safe for humans, though it could be harmful if ingested by small children. It should be applied only in hard-to-reach areas like wall voids, under heavy appliances, and behind cabinets.

You can buy boric acid specifically made to combat roaches. It should have an anticaking compound, since roaches avoid caked powder. One such product is Roach-Kil. It can be easily dispensed into wall voids and cracks (those too large to caulk shut) from a squeeze bottle equipped with a strawlike applicator.

Another solution is insect growth regulator. Hydroprene (Gencor, from Zoecon) is a synthetic version of the roach hormone that regulates growth and development. Hydroprene prevents the juvenile roach from becoming a reproductive adult. After treatment with this chemical, the cockroach population slowly dies out—if no new roaches migrate into the treated areas. This product is considered safe for humans. According to the Bio-Integral Research Center, Gencor can be sprayed on kitchen cabinets, under shelf liners, or on other surfaces where it is not practical to use boric acid. Gencor should not be put on surfaces in contact with food.

Another effective product sold in retail stores is Combat Roach Control Bait Trays (Clorox). These are small plastic enclosures that can be placed in areas where you have seen roaches. The roach enters the enclosure, where it dines on a mix of oatmeal and corn syrup. The mix contains a chemical that blocks the roach's digestion and slowly kills it. Its poisoned remains are eaten by other roaches, which also become sick and die. According to the manufacturer, pets will not be harmed if they eat a poisoned roach.

To determine how your home modification and low-toxicity treatments are working, place roach traps (see Resources) in roach-infested areas or where you suspect the pests are entering your home. This should be done before treatments begin and then several months later.

ANTS

No one likes to see a trail of ants walking over their kitchen floor. Ants do have their good points. Like spiders, they eat fleas, fly larvae, bedbugs, and termites. They decompose plant and animal tissue. Ants don't carry deadly diseases, and most common house varieties don't bite. So their presence is usually only an aesthetic problem.

Home modifications to deter ants are essentially the same as discussed for cockroaches. For a quick kill, spraying ants directly with a solution of soapy water is effective. Other strategies include diatomaceous earth, amorphous silica gel, and boric acid (see Resources), powders that can be blown into cracks and wall voids, which are then sealed with caulk. Inside walls, these chemicals provide years of control (also for cockroaches) and are safely out of the way. Boric acid kills slowly, giving worker ants time to take poison back to the queen and the young. Baits are also effective. One, Drax, contains boric acid and mint jelly as an attractant. Combat sells a bait tray for ants that works on the same principle as its cockroach bait. Baits can be placed where ants have been seen. The manufacturer says they are childproof, and harmless even if a child were to eat the contents. Insect growth regulators for specific ant species are also available. Pharorid is good for difficult-to-control pharaoh ants. The manufacturer suggests using it with boric acid baits. To eliminate an outside ant nest close to your home's foundation, thoroughly drench the nest with insecticidal soap (see Resources). Pyrethrin insecticide from Ecosafe can also be used for this purpose. See chapter 5, "Maintaining a Chemical-Free Yard," for a discussion of pyrethrin insecticides.

Carpenter ants can damage your home by burrowing in rotting wood to form their nests. These ants can be effectively spot-treated with the Electro-gun described earlier in connection with termites.

Fire ants are the most feared species because of their painful sting. They form giant moundlike nests outdoors. When disturbed, fire ants swarm out and attack. Children are particularly vulnerable because they can't outrun the ants. The size of a nest should be reduced so that natural predators like spiders, beetles, and bacteria in the soil can eliminate the rest of the ants. Soapy, boiling water can be poured on the mounds to destroy ants. The insect growth regulator Logic is also effective in controlling ants when sprinkled around the perimeter of a mound.

FLIES

Flies buzzing around your home are a real annoyance, but a preventable one. Make sure your screens fit tight and your doors close snugly. An occasional fly is best controlled by the safe and effective flyswatter or sticky flypaper.

Don't use hanging pest strips, which release pesticide vapors into the air your family breathes. Some strips contain dichlorvos, an animal carcinogen. Dichlorvos has been under review by the EPA for some ten years. In 1987 the EPA, after calculating a cancer risk as high as 1 in 100 for pest strips, proposed that manufacturers place a cancer warning on their product labels. Following a warning in 1988 by *Consumer Reports*, several companies reformulated their products to exclude dichlorvos, but others did not. The EPA has not yet banned products containing dichlorvos and has even backed away from enforcing the cancer labeling until it finishes reviewing the health risks.

HEAD LICE

About 10 percent of the elementary school population is treated every year for this common infestation of the hair and scalp. The lice lay their eggs, or nits, very close to the scalp. The best way to avoid lice is to encourage children not to share combs, hats, or other objects that contact the head. To control an infestation, comb the hair with a special metal lice comb, such as the Innomed Lice comb, available at some drugstores. Hair should be wet and soapy to facilitate movement of the comb. The common plastic comb is useless because the teeth are too far apart to catch and remove eggs. Combing is a painstaking process that needs to be repeated weekly when there is an infestation at school. The good thing is that children often enjoy having their hair combed and don't put up a fuss.

Lindane shampoos, the most common treatment for head lice, pose several problems. Lindane, a pesticide, can pass through the scalp and lodge in fat tissue. The EPA has labeled lindane a possible human carcinogen because of animal data showing it causes liver tumors. Moreover, head lice are often resistant to it. Pyrethrin products like Rid are safer.

Soaps containing certain fatty acids found in coconut and olive oil appear to be effective against young and adult lice and safe to use. Regular shampooing with such soap may prevent an infestation. Wash your child's hair in water as hot as she or he can tolerate; lice are killed at high temperatures.

Contact the National Pediculosis Association for more information on head lice (see Resources).

FLEAS

The most common house flea is the cat flea, which infests both cats and dogs and bites humans as well. Only about 20 percent of the fleas in a home are actually on your pet—the other 80 percent are in carpets, furniture, and pet bedding. To control fleas, launder your pet's bedding regularly. Washing destroys all stages of the flea's life cycle. Vacuuming your home thoroughly and regularly also helps. If the flea population gets out of hand, your upholstery and rugs may require steam cleaning or shampooing. Flea Control for Carpets can be added to carpet cleaning solutions; it prevents fleas from breeding in a carpet for up to a year, according to the manufacturer. Regularly bathing and combing your pet with a special flea comb found at a pet store will help keep fleas under control.

The most common control method is the flea/tick collar containing an organophosphate or carbamate insecticide. The collar emits insecticidal vapors that kill fleas but are also inhaled by animals and adults. Use a collar, if you must, intermittently on your pet, and store it in a tightly sealed container the rest of the time.

An alternative to flea collars is an insecticidal soap like Flea Soap for Dogs and Cats, Flea Stop Flea Spray, and Flea Stop Shampoo. The latter two products contain d-limonene, a citrus-derived insecticide. The shampoo and soap can be used to wash your pet, its bedding, rugs, and other areas in which it hangs out.

The insect growth regulator methoprene can be applied to flea breeding areas like bedding. It prevents maturation of the juvenile flea but will not kill adults. You can powder your animal with the pyrethrin-based Herbal Flea Powder, or you can sprinkle diatomaceous earth from Ecosafe Products on your pet, its bedding, your carpets, furniture, and even the lawn.

CLOTHES MOTHS

The larvae of clothes moths eat fabric. They feed on clothing or upholstery containing wool, feathers, fur, hair, and even blends of wool and synthetic fibers. Moths are interested only in contaminated products that harbor food, drink, sweat, and urine. So the most effective strategy is to keep fabrics clean and stored in well-sealed containers or plastic bags. Regular vacuuming also helps; it minimizes lint, hair, and other organic debris that make your home hospitable to moths. Since moth larvae and cocoons are very fragile, frequent wearing or brushing of your clothes will also help prevent infestation. Larvae also fall off clothes exposed to sunlight. Moths are sensitive to heat; storing your clothes in a hot attic during the summer will kill all stages of the moth life cycle. Cedar chests and closets are not effective in the long run; the wood loses its killing potential a few years after it is cut.

If you follow the suggestions above, there should be no need for chemical insecticides to control moths. The most commonly used chemicals are naphthalene and paradichlorobenzene (PDB). Both are sold in ball, flake, or cake form, designed to sustain a high concentration of chemical vapor over several months. PDB contains the known human carcinogen benzene. It is readily absorbed by the body following inhalation and is present in body fat in individuals who are regularly exposed.

Some dark-skinned individuals are sensitive to naphthalene, which can destroy blood cells. Sensitivity is probably sex-linked, with males being more vulnerable. Naphthalene has also caused violent reactions in children who were dressed in clothes stored with naphthalene mothballs.

Camphor, which is sold in cake form, is another common moth repellent. In the long term, it may be the least toxic of the three moth repellents because it does not concentrate in human fat.

RESOURCES

Products and Services

Biologic, 18056 Springtown Rd., P.O. Box 177, Willow Hill, PA 17271. Telephone: (717) 349-2789.
    Sells nematodes for termite control.
Brody Enterprises, 9 Arlington Pl., Fair Lawn, NJ 07410. Telephone: (800) 458-8727.
    Markets roach traps.
Ecosafe Products, P.O. Box 1177, St. Augustine, FL 32085. Telephone: (800) 274-7387.
    Sells diatomaceous earth for termite, ant, cockroach, and flea control, as well as pyrethrin-based insecticides like Herbal Flea Powder
Etex, 916 S. Casino Center, Las Vegas, NV 89101. Telephone: (800) 543-5651.
    Call for information on the Electrogun for termite control.
Fairfield American, 201 Rte. 17 N., Rutherford, NJ 07070. Telephone: (201) 507-4880.

Sells amorphous silica gel products Dri-die 67 and Drione for treatment of termites, cockroaches, and ants.

Farnam Pet Products, P.O. Box 34820, Phoenix, AZ 85067. Telephone: (800) 343-9911.

Sells the citrus oil products Flea Stop Flea Spray and Flea Stop Shampoo for flea control.

Isothermics, P.O. Box 18703, Anaheim, CA 92817. Telephone: (714) 778-1396.

Markets sand barriers and heat treatment for termites.

Paragon, P.O. Box 17167, Memphis, TN 38187. Telephone: (800) 238-9254.

Sells Logic, an insect growth regulator for fire ants.

R-Value/West, 150 N. Santa Anita Ave., Suite 300, Arcadia, CA 91006. Telephone: (818) 798-4000.

Sells boric acid products to treat cockroach, termite, and ant problems, Flea Control for Carpets, and Drax ant traps.

Safer, 189 Wells Ave., Newton, MA 02159. Telephone: (800) 423-7544.

Markets insecticidal soap, including Flea Soap for Dogs and Cats, and the pyrethrin-based insecticide Flea and Tick Spray.

Zoecon, 12005 Ford Rd., Suite 800, Dallas, TX 75234. Telephone: (800) 527-0512.

Sells insect growth regulators: Gencor for cockroaches, Pharorid for pharaoh ants, and Precor for fleas.

Sources of Information

The Bio-Integral Resource Center, P.O. Box 7414, Berkeley, CA 94707. Telephone: (415) 524-2567.

Write away for the following IPM publications: "Ants," "Cockroaches," "Fabric and Paper Pests," "Flies," "IPM for the Cat Flea," "IPM for Head Lice," and "Termites and Other Wood-Damaging Pests."

EPA Public Information Center, 401 M St. SW, PM 211B, Washington, D.C. 20460. Telephone: (202) 475-7751.

Write for "Termiticides: Consumer Information," February 1988 (OPA-87-014).

National Pediculosis Association, P.O. Box 149, Newton, MA 02161. Telephone: (617) 449-NITS.

Provides consultation on head lice problems.

National Pesticide Telecommunications Network. Telephone: (800) 858-7378.

Provides information on pesticide toxicity as well as the proper use and disposal of pesticides.

Pest Control You Can Live With. Debra Graff. Sterling, VA: Earth Stewardship Press, 1990.

A handbook on nontoxic household pest control ($5.25). Write to the National Center for Environmental Health Strategies, 1100 Rural Ave., Voorhees, NJ 08043, or phone (609) 429-5358.

CHAPTER 18

Disposing of Hazardous Household Products

The Problem. Families are exposed again to toxic household products after their disposal when they contaminate air, soil, and water.
The Risk to Kids. Children are more vulnerable to contaminated air, soil, and water for physiological and behavioral reasons.
What to Do. Parents can reduce disposal hazards by selective purchasing, recycling, and participating in hazardous waste collections.

Household chemicals pollute not only the indoor environment but also the outdoors. Soil becomes contaminated when people pour toxics into it or send them with the garbage to the landfill. Air gets contaminated when toxics are incinerated or allowed to evaporate. Water is contaminated when toxics are flushed into rural septic tanks or municipal water-treatment systems, and when the leachate from landfills seeps into groundwater.

Discarded paint is a good example of how household waste contaminates the environment. Most families don't worry about throwing half a can of paint down the drain or in the trash now and then. But it turns out that Americans use 3 million gallons of paint daily—that's a lot of leftover paint being tossed away. As we have seen, evaporating organic solvents from oil-based paint contribute to ozone air pollution. If the paint is sent in trash to a landfill, it can leak out of the can and be washed by rain into the ground, threatening soil and waterways. If paint is flushed down the toilet, poured down the drain, or dumped into a sewer drain, it will not be detoxified by conventional wastewater treatment. Paint toxics contaminate the sewage sludge formed during water purification, and when released with wastewater they can contaminate the aquatic food chain. If toxics enter a rural septic tank system, they kill the organisms that degrade waste. They also pass unchanged into the drainage field and on to local groundwater. Many rural homes use this groundwater for drinking.

THE MOST COMMONLY DISPOSED TOXICS

The average American home generates 15 pounds of hazardous waste yearly. Based on an analysis of collection programs, the most common waste is paint and related products like thinner. Other common waste includes dirty motor oil, household cleaners, pesticides, antifreeze, batteries, and hobby chemicals.

HOW CAN HAZARDOUS WASTE BE REDUCED?

The ideal way to reduce hazardous waste in the home is to stop buying hazardous products. Since some of these products are hard to replace and it takes time to change consumer habits, other measures are also necessary: recycling, returning hazardous products to manufacturers, sharing and donating products, and using a hazardous waste collection service. Families should also support legislation to reduce hazardous waste and to make waste treatment facilities operate safely.

Source Reduction

Source reduction—not buying hazardous products—is the easiest and cleanest way to reduce such waste. No choices have to be made about whether to bury, burn, or recycle. Families can choose alternative products like those mentioned in chapter 16. A large stride toward halting pollution will be made if millions of people change their habits at home.

Recycling

At present, there are a lot of toxic products that families can't avoid buying. Examples are car batteries, motor oil, and paint. The best way to dispose of them is to recycle. To find out about recycling toxics, contact the Environmental Hazards Management Institute (EHMI) (see Resources). For example, auto products like batteries, dirty motor oil, and brake and transmission fluid can be brought to a service station, reclamation center, or a hazardous waste collection site. People who change motor oil themselves can buy an oil recycling kit. The kit catches the dirty oil in a container, which is then taken to a service station. If everyone in a city of 100,000 people managed to recycle motor oil, it would save over 3 tons of dirty motor oil a month. Manufacturers welcome it because much less dirty oil is needed than virgin oil to make new oil.

Another good candidate for recycling is latex paint, which can be mixed and reused. A latex paint recycling program in Seattle mixes certain latex paints (no dark colors and no orange or yellow colors, which might contain lead) to produce a color called Seattle beige. Much of it goes to schools and hospitals. If you must discard a latex paint, let it air dry and discard the container in the trash. Oil-based paint, which is more toxic than latex or water-based paint, is not as good a candidate for recycling. However, it can be used up, or leftovers exchanged or taken to a hazardous waste collection site. The city of Santa Monica, California, has recirculated more than 600 gallons of paint through a paint exchange.

Several household items not considered hazardous but which can become so if incinerated, for instance, newspapers, glass, aluminum and steel cans, and plastic bottles, are now being recycled in many communities. Discarded glass and aluminum are recycled to make new glass and aluminum containers. Discarded plastics used to make things like benches and fencing. If your community has no recycling program, you can get relevant information from the Environmental Defense Fund or contact the National Recycling Coalition (see Resources).

Teaching Children about Recycling

Parents can support recycling activities in their community and teach their children to participate. Find out exactly how your community recycles by calling the local sanitation or public works department. Successful recycling starts with shopping. Start by bringing along reusable shopping bags. At the store, have your children search for items that can be recycled, and buy larger sizes with more efficient packaging. Look for recycled-paper packaging and avoid products with excessive or single-helping packaging.

Teach your children what kinds of paper, glass, metals, and plastics are currently accepted for recycling; for example, newspapers, corrugated cardboard, and computer paper are accepted, (shiny colored paper and paper with food on it are not recycled). All glass except ovenproof is recyclable, as long as metal caps and rings are removed. Plastic containers are beginning to appear with a code number stamped on the bottom. Some areas accept all numbers, others only code 1 and 2 (polyethylene terephthalate and high-density polyethylene), which usually appear on soda bottles and milk jugs, respectively. Code 3 (vinyl) appears on shampoo bottles, code 4 (low density

polyethylene) on squeeze bottles, code 5 (polypropylene) on ketchup bottles, and code 6 on polystyrene packaging. You can distinguish aluminum
from steel cans with a magnet, as only steel cans are magnetic. Remove
labels from cans.

Returning Hazardous Products to the Manufacturer

A few hazardous products should be returned to the manufacturer. One is
any pesticide containing banned ingredients. Another is the ionization
smoke detector, which emits alpha-ionizing radiation. The Nuclear Regulatory Commission says there is little or no risk associated with the use of
these detectors. However, millions are manufactured, used, and discarded
yearly, and scientists note there is no threshold for ionizing-radiation safety.
You can avoid ionizing smoke detectors if you are willing to pay about $10
more for a photoelectric smoke detector (see Resources). The photoelectric
detector responds faster to slow, smoldering fires like those started in a
mattress by a cigarette.

Smart Use and Smart Storage

Families can minimize the effect of hazardous products by developing new
habits of using and storing them. Read the label on a product before buying
it and make sure it does the job you need it for. Also read any warning and
the instructions for use and disposal. Buy the minimum amount needed for
the job, and do your best to use the product up. If you have some left over,
don't throw it out. Instead, share it with a friend or neighbor, donate it to a
school, park district, church, or community theater group, or check to see if
it can go to a recycling or exchange program.

If your product must be stored, leave it in the original container, if
possible with the label and instructions intact. Don't combine or repackage
products, or put them in food containers. Cap the container tightly and
store outside the house in a dry, well-ventilated area that is protected and
not accessible to children.

Hazardous Waste Collection Programs

To find out about a hazardous waste collection program in your area, call
your waste management company, local health department, or state EPA.
Some states are instituting year-round collection centers, others sponsor
annual pickups in various regions of the state. Waste should be identified,
packaged, and labeled by trained personnel. Most programs are run by
public agencies that have responded to pressure from local citizen groups.
About 2,000 programs were established in forty-five states between 1981

and 1990. But thus far they have attracted only 5 percent of the population they are designed to serve.

If you want to organize a collection program in your area, write away for the Household Hazardous Waste Information Kit from the League of Women Voters of Massachusetts and the quarterly newsletter *Household Hazardous Waste Management News* (see Resources). One member of the League of Women Voters who set up a successful program was Joan Dotson of Redlands, California. In 1985, she and her local chapter staged a clean-up day for the town that evolved first into an annual event and then into a permanent collection site funded at the reasonable cost of 16 cents a month, added to residential trash collection bills. The group spent nine months preparing for the first collection, securing from city, county, and private donors a site, funds for advertising, educational and promotional materials, and the services of a hazardous waste hauler.

Annual collections mean that families have to store household toxics for a considerable time. Use the storage guidelines mentioned above, and when the day arrives, be careful how you transport your waste to the site. Keep children and pets out of the way, and don't smoke while handling chemicals. Seal containers tightly; wrap leaky containers in newspaper. Separately box flammables, corrosives, and poisons. Load them in the car as far from passengers as possible.

Collection programs don't accept every kind of waste. They don't take unidentified, nonhazardous (this may include latex paint), radioactive, explosive, infectious and medical, or commercial waste. Moreover, while collection programs perform a wonderful service by taking hazardous waste off the hands of consumers, they have no perfect way to dispose of it. Waste is sorted and transported mostly by truck to facilities sometimes out of state. According to the Institute of Chemical Waste Management, about 1 percent of waste is recycled; 72 percent is treated chemically or biologically to reduce toxicity; and 1 percent is incinerated. The residue, plus any material that cannot be treated, is buried in landfills. So even with the most careful waste collection, sorting, and treatment, hazardous waste residue still ends up in the soil. And since no landfill has been designed that can guarantee against leakage, waste threatens to migrate farther into soil and into water. This emphasizes the need for source reduction and new technologies to detoxify hazardous waste.

Hazardous Waste Legislation

Most families would not welcome a new incinerator or landfill in their area. If one is to be built in your community, you can petition your state and federal representatives for state-of-the-art hazardous emission control devices. New incinerators should be equipped with stack scrubbers to detoxify fumes, and hazardous ash should be collected for special disposal. Landfills should have the best technology available, including leachate collection systems, to keep toxics out of soil and groundwater. Groundwater monitoring programs should operate at landfills.

Families would be wise to review the new bills in Congress and the Senate proposed to reauthorize the Resource Conservation and Recovery Act (RCRA), which expired in 1988. The RCRA was passed in 1976 and governs solid and hazardous waste. Contact your representatives if you support the bills. One House bill (HR 3735), introduced by Senator Thomas Luken (Democrat, Ohio), would require the EPA every five years to identify five of the most common toxic constituents of municipal waste and to consider imposing one of the following: a ban on their use, a ban on landfilling or incineration, special management standards, or substitutes. This same bill calls for recycling lead-acid and mercury batteries and for an EPA study into means for recycling household dry-cell batteries and methods for their disposal. Finally, the bill encourages the recycling of oil.

You can find out how many of various hazardous materials manufactured or used outside your state have been shipped into your state for disposal. Consult the Toxics Release Inventory (see Resources). A National Library of Medicine computerized database, the Hazardous Substances Data Bank, gives information on chemical toxicity, emergency handling procedures, environmental fate, human exposure, detection methods, and regulatory requirements.

RESOURCES

Products and Services

Environmental Hazards Management Institute (EHMI), P.O. Box 932, 10 Newmarket Rd., Durham, NH 03824. Telephone: (603) 868-1496.
> Sells the Household Hazardous Waste Wheel, the Recycling Wheel, and the Daily Recycler Wheel.

Laidlaw Environmental Services, P.O. Box 210799, Columbia, SC 29221. Wats: (800) 845-1019. Telephone: (803) 798-2993. Fax: (803) 798-3660.
> Assists communities in planning hazardous waste collection programs. Write for information.

League of Women Voters of Massachusetts, 8 Winter St., Boston, MA 02108. Telephone: (617) 357-8380.
> Sells for $25 the Household Hazardous Waste Information Kit. Will also sell

or rent by the week instructional videos, 16mm film, and slide shows on the subject.

Pittway, 780 McClure Rd., Aurora, IL 60504. Telephone: (708) 851-7330.

Sells the BRK model 2001 photoelectric smoke detector and takes back discarded ionizing smoke detectors.

Toxicology Information Program, National Library of Medicine, 8600 Rockville Pike, Bethesda, MD 20894. Telephone: (301) 496-6531.

Provides computer access to the annual TRI as well as a list of toxic emissions released by industry to air, water, and land.

Sources of Information

Coming Full Circle: Successful Recycling Today.

To order this publication ($12), write the Environmental Defense Fund, 257 Park Ave. S, New York, NY 10010. Telephone: (212) 505-2100.

Garbage.

A magazine devoted to issues like recycling. Write to P.O. Box 51647, Boulder, CO 80321. Telephone: (800) 274-9909.

Household Hazardous Waste Management News.

A free quarterly newsletter. Write to Dana Duxbury and Associates, Waste Watch Center, 16 Haverhill St., Andover, MA 01810. Telephone: (508) 470-3044.

National Recycling Coalition, 1101 30th St. NW, Suite 305, Washington, D.C. 20007. Telephone: (202) 625-6406.

Send a self-addressed, stamped envelope for a packet of information for communities interested in recycling.

Office of Solid Waste and Emergency Response (WH-562), U.S. EPA, 401 M St. SW, Washington, D.C. 20460. Telephone: (800) 424-9346 and (703) 920-9810.

Has a list of hazardous waste experts in each state. Also provides information on landfill regulations and hazardous waste disposal.

Appendix:
EPA Regional Offices

REGIONAL OFFICES	STATES AND TERRITORIES COVERED
EPA Region I Room 2203 John F. Kennedy Federal Building Boston, MA 02203 (617) 565-3715	Connecticut, Maine, Massachusetts, New Hampshire, Rhode Island, Vermont
EPA Region II 26 Federal Plaza New York, NY 10278 (212) 264-2515	New Jersey, New York, Puerto Rico, Virgin Islands
EPA Region III 841 Chestnut St. Philadelphia, PA 19107 (215) 597-9800	Delaware, District of Columbia, Maryland, Pennsylvania, Virginia, West Virginia
EPA Region IV 345 Courtland St. NE Atlanta, GA 30365 (800) 282-0289 in GA; (800) 241-1754 in other Region IV states	Alabama, Florida, Georgia, Kentucky, Mississippi, North Carolina, South Carolina, Tennessee

EPA Region V   Illinois, Indiana, Michigan, Minnesota, Ohio,
230 South Dearborn St.   Wisconsin
Chicago, IL 60604
(800) 572-2515 in IL; (800)
621-8431 in other Region V
states

EPA Region VI   Arkansas, Louisiana, New Mexico, Oklahoma,
12th Floor, Suite 1200   Texas
1445 Ross Ave.
Dallas, TX 75270
(214) 655-2200

EPA Region VII   Iowa, Kansas, Missouri, Nebraska
726 Minnesota Ave.
Kansas City, KS 66101
(913) 551-7003

EPA Region VIII   Colorado, Montana, North Dakota, South Dakota,
999 18th St.   Utah, Wyoming
Denver, CO 80202-2405
(800) 759-4372

EPA Region IX   Arizona, California, Hawaii, Nevada, American
215 Freemont St.   Samoa, Trust Territories of the Pacific, Guam,
San Francisco, CA 94105   Northern Marianas
(415) 556-6608

EPA Region X   Alaska, Idaho, Oregon, Washington
1200 Sixth Ave.
Seattle, WA 98101
(206) 442-5810

Note: States have various environmental agencies—such as a state EPA, Department of Natural Resources, or a Department of Health—that provide information to the public. Because air quality issues are divided among many state agencies, a "Directory of State Indoor Air Contacts" is now available. Contact your regional EPA office if you are interested.

Selected Bibliography

Ames, B. N., et al. "Ranking Possible Carcinogenic Hazards." *Science* 236 (1987): 271–78.

Ames, B. N., and L. S. Gold. "Too Many Rodent Carcinogens: Mitogenesis Increases Mutagenesis." *Science* 249 (1990): 970–71.

Archer, M. C. "Hazards of Nitrate, Nitrite and N-Nitorso Compounds in Human Nutrition." in *Nutritional Toxicology.* Vol. 1, edited by J. N. Hathcock. New York: Academic Press, 1982.

Bean, N. H., et al. "Foodborne Disease Outbreaks: 5-Year Summary, 1983–1987." *Morbidity and Mortality Weekly Report* 39, no. SS-1 (1990): 15–58.

Bierman, C. W., and D. S. Pearlman. *Allergic Diseases from Infancy to Adulthood.* 2d ed. Philadelphia: W. B. Saunders, 1988.

Branen, A. L., et al. *Food Additives.* New York: Marcel Dekker, 1990.

Brunner, C. R. *Hazardous Air Emissions from Incineration.* New York: Chapman and Hall, 1986.

Calabrese, E. J. *Age and Susceptibility to Toxic Substances.* New York: John Wiley, 1986.

Check, W. A. "Is Anyone Safe from the Hazards of Pollution?" *American College of Physicians Observer* 10 (1990).

"Childhood Lead Poisoning, United States: Report to the Congress by the Agency for Toxic Substances and Disease Registry." *Morbidity and Mortality Weekly Report* 37 (August 19, 1988): 481–485.

Congress of the United States, Office of Technology Assessment. *Neurotoxicity: Identifying and Controlling Poisons of the Nervous System.* OTA-BA-436. Washington, D.C.: GPO, 1990.

Davis, J. M., and P. L. Svendsgaard. "Lead and Child Development." *Nature* 329 (September 1987): 297–300.

Detels, R., et al. "The UCLA Population Studies of Chronic Obstructive Respiratory Disease." *American Journal of Epidemiology* 109 (1979): 33–58.

Edwards, D. D. "ELF: The Current Controversy." *Science News* 131 (1987): 107–9.

Elkan, R., and C. Carroll. "Latest Weapons against Melanoma Focus on Education, Computer Analysis." *American College of Physicians Observer* 10 (1990).

Environment '90: The Legislative Agenda. Washington, D.C.: Congressional Quarterly, 1990.

Feigin, R. D., and J. D. Cherry. *Textbook of Pediatric Infectious Diseases.* 2d ed. Philadelphia: Saunders, 1987.

Fenske, R. A., et al. "Potential Exposure and Health Risks of Infants Following Indoor Residential Pesticide Applications." *American Journal of Public Health* 80 (1990): 689–93.

Fielding, J. E., and K. J. Phenow. "Health Effects of Involuntary Smoking." *The New England Journal of Medicine* 319 (1988): 1452–60.

Fisk, W. J., et al. *Indoor Air Quality Control Techniques: Radon, Formaldehyde, Combustion Products.* Park Ridge, NJ: Noyes Data, 1987.

Gabler, R., and the editors of *Consumer Reports Books. Is Your Water Safe to Drink?* New York: Consumers Union, 1988.

Gammage, R. B., and S. V. Kaye. *Indoor Air and Human Health.* Chelsea, MI: Lewis, 1985.

Hallenbeck, W. H., and K. M. Cunningham-Burns. *Pesticides and Human Health.* New York: Springer-Verlag, 1985.

The Healthy House Catalog. Cleveland: Environmental Health Watch and the Housing Resource Center, 1990.

Hurwitz, S. "The Sun and Sunscreen Protection: Recommendations for Children." *Journal of Dermatologic Surgery and Oncology* 14 (1988): 657–60.

Immerman, F. W., and J. C. Schaum. *Nonoccupational Pesticide Exposure Study (NOPES).* Washington, D.C.: U.S. EPA, 1990.

Janerich, D. T., et al. "Lung Cancer and Exposure to Tobacco Smoke in the Household." *New England Journal of Medicine* 323 (1990): 632–36.

Kane, D. N. *Environmental Hazards to Young Children.* Phoenix: Oryx Press, 1985.

Kirkpatrick, D. C., et al. "Food Packaging Materials: Health Implications." In *Nutritional Toxicology.* Vol 3. Edited by J. N. Hathcock. New York: Academic Press, 1989.

Lansdown, R., and W. Yule, eds. *The Lead Debate: The Environment, Toxicology, and Child Health.* London: Croom Helm, 1986.

Lowengart, R., et al. "Childhood Leukemia and Parents' Occupational and Home Exposures." *Journal of the National Cancer Institute* 79 (1987): 39–46.

Monastersky, R. "Antarctic Ozone Bottoms at Record Low." Science News 138 (1990): 228.

Mossman, B. T., et al. "Asbestos: Scientific Developments and Implications for Public Policy." *Science* 247 (1990) 294–302.

Needleman, H. L., et al. "The Long-Term Effects of Exposure to Low Doses of Lead in Childhood: An 11-Year Follow-up Report." *New England Journal of Medicine* 322 (1990): 83–88.

Nero, A. V. "Controlling Indoor Air Pollution." *Scientific American* (May 1988): 42–48.

Olkowski, W. "Safe Mosquito Management." *Common Sense Pest Control Quarterly* 3, no. 2 (1987): 5–16.

Olkowski, W., et al. "Managing Ticks—the Least Toxic Way." *Common Sense Pest Control Quarterly* 6, no. 2 (1990): 4–25.

Olney, J. W. "Excitotoxic Food Additives: Functional Teratological Aspects." *Progress in Brain Research* 73 (1988): 283–94.

Pitts, D. G. "Ultraviolet-Absorbing Spectacle Lenses, Contact Lenses, and Intraocular Lenses." *Optometry and Vision Science* 67 (1990): 435–40.

Pool, R. "Electromagnetic Fields: The Biological Evidence." *Science* 249 (1990): 1378–81.

———. "Flying Blind: The Making of EMF Policy." *Science* 250 (1990): 23–25.

———. "Is There an EMF-Cancer Connection?" *Science* 249 (1990): 1096–98.

Pope, C. A. "Respiratory Disease Associated with Community Air Pollution and a Steel Mill, Utah VAlley." *American Journal of Public Health* 79 (1989): 623–28.

Smith, K. R. "Air Pollution—Assessing Total Exposure in the United States." *Environment* 30 (1988).

Spektor, D. M., et al. "Effects of Ambient Ozone on Respiratory Function in Active, Normal Children." *American Review of Respiratory Diseases* 137 (1988): 312–20.

U.S. General Accounting Office. *Lawn Care Pesticides: Risks Remain Uncertain while Prohibited Safety Claims Continue.* March, 1990.

Wallace, L. A. *The Total Exposure Assessment Methodology (TEAM) Study: Summary and Analysis.* Vol. 1. Washington, D. C.: EPA, 1987.

Ware, J. H., et al. "Passive Smoking, Gas Cooking and Respiratory Health of Children Living in Six Cities." *American Review of Respiratory Disease* 129 (1984): 366–74.

Weiss, K. B., and D. K. Wagener. "Changing Patterns of Asthma Mortality." *Journal of the American Medical Association* 264 (1990): 1683–87.

Index

"Potential Exposures and Health Risks of
Infants Following Indoor Residential
Pesticide Applications," 189
Pottery, lead in, 120–21
Poultry, 5, 97
bacteria in, 98–102
grilled, 101–102
irradiated, 101
pesticides in, 95
safe handling of, 99–101
Power lines, 165, 166, 168–69
Predators, natural, 47
Pregnant women, 106, 115, 135, 147, 169,
170
dangers to the fetus, 10, 35, 147
Premature aging of the skin, 77, 79
Preservatives, *see* Food additives
Pressed-wood products, 148, 149, 151, 152
Processed foods, additives in, *see* Food
additives
Procter and Gamble, 117
Produce, *see* Fruits and vegetables, pesticides
in
Pronamide, 46
Propoxur, 189
Pyrethrin, 53, 194, 195

Quaker Oats, 101

Radiation exposure, *see specific sources of
radiation exposure, e.g.* Radon
Radium, 130
Radon, 5, 6, 18–26, 153
areas with high levels of, 19–20
creation of, 19
in drinking water, 20, 25, 129–30,
135–36
increased vulnerability of children to, 19
levels of, 23
lung cancer and exposure to, 19, 23, 24,
129–30
methods for eliminating, 24–26
radon daughters, 19
smoking and risk from, 24
testing for, 21–22, 25
ways it enters the home, 20
when to take action, 23–24
*Radon: A Homeowner's Guide to Detection and
Control* (Cohen), 20
"Radon Reduction Methods: A Homeowner's
Guide," 25
Ralston Purina, 101
Reader's Digest, 74
Recycling, 50, 199–201
Red Dye No. 3, 116–17
Refrigeration of food, 99

Refrigerators, 160, 161
coolant, 75, 76, 77, 82
Remote Access Chemical Hazards Electronic
Library (RACHEL), 178
"Removal of Radon from Household Water,"
135
Renovation, lead dust during, 14–15
Repellent plants, 47–48
"Residential Air-Cleaning Devices: A
Summary of Available Information,"
152, 162
Resource Conservation and Recovery Act, 204
Respiratory diseases, 4, 65, 67–68, 143,
156, 161
see also Asthma; Emphysema
Retardation, 9, 115
Returning hazardous products, 202
Reuben, Dr. David, 74
Reverse-osmosis water purification systems,
134, 135
Rigel, Dr. Darrell, 76
Roach-Kil, 193
Rocky Mountain spotted fever, 57, 58
*Rodale's Garden Insect, Disease, and Weed
Identification Guide* (Smith and Carr), 51
Rugs, *see* Carpeting
Ryan, Kevin, 44, 45

Saccharin, 113
Safe Computing, 170
Safe Drinking Water Act, 124, 130
Salicylates, 79
Salmonella, 98, 101, 108
Salmonellosis, 98–99
Sargeant Art, 36
"Save Our Streams," 137
Savitz, David, 166
Schultz, Warren, 49
Scombrotoxin, 105
Sediment contamination, surface water, 125
Self-tanning lotions, 80
Septic systems, 125, 126, 136
Shigella, 99
Sick-building syndrome, 148
Sigma Designs, 170, 171
Silica, 34, 38
Simplesse, 117
Sinan, 185
Skin cancer, 74–79
Slesin, Louis, 169
Smog alerts, 70
Smoke detectors, 202
Smoking, 24, 146–48
Sodium hydroxide, 178–79
Sodium hypochlorite, 178, 179
Soil, 48
Solvents, toxic, 34, 179–80

About the Authors

JOYCE SCHOEMAKER received a Ph.D. in microbiology from Jefferson Medical College in Philadelphia. She has taught and done research in molecular biology at the University of Chicago and has held positions in research and management at several biotechnology companies. She is the author of numerous research publications. Dr. Schoemaker lives in the Chicago area with her husband Paul and young twins, Kimberly and Paul.

CHARITY VITALE has a Ph.D. in biology from Georgetown University. She taught biology at St Joseph's College in Philadelphia. She has done biophysical research at the National Biomedical Research Foundation in Washington, D.C., and is the author of several scientific publications. Dr. Vitale lives near Chicago with husband David and their young children, Laura and Peter.

Both authors have been actively involved with environmental issues in their communities.

Also Available from Island Press

Ancient Forests of the Pacific Northwest
By Elliot A. Norse

Balancing on the Brink of Extinction: The Endangered Species Act and Lessons for the Future
Edited by Kathryn A. Kohm

Better Trout Habitat: A Guide to Stream Restoration and Management
By Christopher J. Hunter

Beyond 40 Percent: Record-Setting Recycling and Composting Programs
The Institute for Local Self-Reliance

The Challenge of Global Warming
Edited by Dean Edwin Abrahamson

Coastal Alert: Ecosystems, Energy, and Offshore Oil Drilling
By Dwight Holing

The Complete Guide to Environmental Careers
The CEIP Fund

Economics of Protected Areas
By John A. Dixon and Paul B. Sherman

Environmental Agenda for the Future
Edited by Robert Cahn

Environmental Disputes: Community Involvement in Conflict Resolution
By James E. Crowfoot and Julia Wondolleck

Forests and Forestry in China: Changing Patterns of Resource Development
By S. D. Richardson

The Global Citizen
By Donella Meadows

Hazardous Waste from Small Quantity Generators
By Seymour I. Schwartz and Wendy B. Pratt

For a complete catalog of Island Press publications, please write:
Island Press, Box 7, Covelo CA 95428, or call: 1-800-828-1302